Kishwaukee College Library
2100 Malta Road
Malta, IL 60150-

D1568016

PEACE OF
MIND IN
EARTHQUAKE
COUNTRY

PEACE OF MIND IN EARTHQUAKE COUNTRY

How to Save Your Home, Business, and Life

3rd Edition, Updated and Expanded

Peter I. Yanev and Andrew C. T. Thompson
Preface by Dianne Feinstein, United States Senator

CHRONICLE BOOKS
SAN FRANCISCO

Copyright © 2008 by Peter I. Yanev and Andrew C. T. Thompson. All rights reserved. No part of this book may be reproduced in any form without written permission from the publisher.

Pages 190–191 constitute a continuation of the copyright page.

Library of Congress Cataloging-in-Publication Data:

Yanev, Peter I., 1946–
 Peace of mind in earthquake country : how to save your home, business, and life / Peter I. Yanev, Andrew C. T. Thompson.
   p. cm.
   Includes bibliographical references and index.
  ISBN 978-0-8118-6183-0
  1. Buildings—Earthquake effects. 2. Dwellings—West (U.S.) I. Thompson, Andrew C. T. II. Title.

TH1095.Y36 2008
693.8'52—dc22

2008005688

Manufactured in Canada.
Designed by Amanda Poray

All maps, diagrams, and other information are approximate and for illustrative purposes only. Neither the publisher nor authors make any warranties concerning the information in this book, which should not be considered as a substitute for professional consultation regarding the situation or construction of any particular building.

10 9 8 7 6 5 4 3 2 1

Chronicle Books LLC
680 Second Street
San Francisco, California 94107
www.chroniclebooks.com

ACKNOWLEDGMENTS

We would like to thank all those who have contributed to the first two editions of the book. We are additionally indebted to the following people for helping to make this edition a reality:

J. Luke Blair at the United States Geological Survey for creating most of the maps in the book and for writing Appendix C

Dan Clark of ReadyDemo for creating the new line art

Chronicle Books, in particular Jay Peter Salvas, Matt Robinson, Sarah Malarkey, Jennifer Kong, Ann Spradlin, Doug Ogan, Molly Jones, Megan Geer, and Andrew Schapiro

Natalie Crosman for providing much editorial assistance and bringing the second edition into digital format

David Strykowski

Arup, in particular Jim Quiter, Aidan Hughes, Mike Kaye, and Michael Willford

Andrew S. Whittaker

Kit Miyamoto of Miyamoto International

Tom K. Chan of Global Risk Miyamoto

Douglas O. Frazier

Kay, Andrew, Alex, and Abigail

TO FRANK AND MIMI BARON
—PETER

TO MOM AND DAD
—ANDY

CONTENTS

Preface 9

Foreword 10

Introduction: Earthquake Risk Management 101 12

Chapter 1 A Primer on Earthquakes: How Earthquakes Are Caused and Measured 21

Chapter 2 The Hazards of Faults and Faulting 39

Chapter 3 Rock, Soils, and Foundations 51

Chapter 4 Other Hazards to Your Property 69

Chapter 5 The Principles of Earthquake Resistance in Buildings 75

Chapter 6 The Best and Worst Types of Construction for Earthquake Resistance 81

Chapter 7 Structural and Architectural Design for Earthquake Resistance 111

Chapter 8 How to Minimize Interior, Utility, and Equipment Damage 145

Chapter 9 Earthquake Insurance 161

Chapter 10 What to Do Before, During, and After an Earthquake 167

Appendix A. Earthquake Hazard Map of the United States 174

Appendix B. Earthquake Hazard Map of the World 176

Appendix C. Earthquake Information in the Digital World by J. Luke Blair, USGS 178

Index 184

Photographic, Map, and Illustration Credits 190

PREFACE

U.S. Senator Dianne Feinstein

In the past few decades, California, like other earthquake-prone states, has made great strides toward earthquake safety. But we still have a long way to go; there are still a staggering number of unsafe old buildings that have not yet been fixed. The mainland United States has yet to experience a major earthquake in modern times. A repeat of the 1906 San Francisco, the 1868 Hayward Fault, or the 1812 New Madrid earthquakes, for example, would cause severe damage.

The problem is not a lack of science and engineering knowledge, as Yanev and Thompson attest in their well-written and informative book. The problem has more to do with the implementation of sound engineering design, and a general lack of education about earthquakes and what we can do to protect ourselves, family, property, and businesses.

This book meets that education challenge. It fills a void in the literature between general preparedness tips and the complex books and papers that fill engineering libraries.

I encourage everyone living in California, and in other earthquake-prone states, to educate and prepare themselves for earthquakes. Reading and acting on the guidance in this book are good first steps.

FOREWORD

by Peter I. Yanev

When the first edition of this book was published in 1974, and I was all of twenty-eight years old, I had personally observed the effects of a handful of real earthquakes. My book was based mostly on hundreds of technical publications and the observations, photographs, and research of many earthquake and structural engineers, geotechnical engineers, seismologists, and academics. Some of the most experienced scientists and engineers at the U.S. Geological Survey, the California Geological Survey, the Earthquake Engineering Research Institute, and the Structural Engineers Association of California gave me their support and unlimited access to their databases and files. Before *Peace of Mind in Earthquake Country*, there was no earthquake-safety publication aimed at homeowners and renters that addressed the issue of what they could do to protect themselves, their families, and their homes in a strong earthquake.

The book, much to the delight of my family and my publisher, Chronicle Books, and to my great surprise, became a best seller. A few days after its publication, a colleague stopped by my desk to tell me that he had observed several people reading the book in his Bay Area Rapid Transit metro car on his commute into San Francisco that morning. Apparently, the book had filled a gap, much as I had hoped.

A few years later, as an experienced earthquake engineer, I realized that the business world lacked the guidance that my book had so easily provided to the residents of earthquake country. Some of the world's best-known companies were based in the earthquake-prone San Francisco Bay Area, but they knew essentially nothing about the risks they faced from a future earthquake. Yet they were very curious. Gary Toms, risk manager of Intel, called me one day in 1980 to invite me to speak about earthquakes to a group of Intel executives. Gary was concerned with the risks his company faced and had the responsibility to minimize or insure against those risks to an acceptable level. In 1980, earthquakes were a risk that all companies worried about but did not know how to address—they simply bought earthquake insurance (if they could get it), much as many companies still do today. So, Intel became my first corporate client. A little while later, Gary invited me to speak to a curiously named group—the Risk and Insurance Management Society (better known as RIMS). RIMS met their first earthquake engineer (me), and I met many future corporate clients. Evaluating the safety of a high-tech plant and strengthening it for earthquakes is not that different from evaluating and strengthening a house. It is a much bigger effort—but the concepts are identical.

A few months later, in 1981, my friend Douglas O. Frazier, who was then a developer and a businessman, and I started our own engineering company, EQE International. The idea was simple: help the business and industry communities to assess their earthquake risks and then help to fix them. By January 1, 2000, when we sold our company, we had a worldwide operation with about 750 people and many offices throughout the United States, Europe, Asia, and elsewhere, with headquarters in my beloved San Francisco Bay Area. Again we had a best seller, this time with our company, especially within the RIMS community but also around the world. From earthquakes we moved into other catastrophic risks—including hurricanes, typhoons, windstorms, floods, conflagrations, industrial accidents, terrorism, and financial risks related to all of the above.

Building on my interaction with risk managers and Doug's business experience, in 1981 we developed a new concept in structural, earthquake, and risk engineering—what we came to call our Earthquake Risk Reduction Program for corporations, which is also applicable for homes and small businesses. It is a simple three-phase program designed to (1) assess earthquake risk, (2) decide how to manage or reduce that risk, and (3) implement decisions such as strengthening buildings and/or buying insurance. This was the idea that built our company into one of the largest engineering companies in the United States, the largest structural and earthquake engineering company in California, and the largest risk management engineering company in the world.

Following the 1989 Loma Prieta (San Francisco) earthquake, I updated *Peace of Mind in Earthquake Country* to include the lessons that engineers had learned in the fifteen years since its original publication. Much has happened to our understanding of earthquakes and their effects, but our best laboratory was, and still is, the earthquakes themselves. Engineers can directly observe their effects on structures and then replicate them in the laboratory and on the computer. We have found out, for example, that ground shaking can be much stronger than we had observed before, that modern steel-framed buildings can be severely damaged, that earthquakes can happen in unexpected places, and that some modern design concepts work well in earthquakes and others do not.

I have visited or sent investigating teams of engineers to the sites of more than one hundred earthquakes throughout the world, and the results of these investigations were extensively reported in the technical and business media. Hundreds of papers and research reports were also published, and my library now includes many tens of thousands of photographs of earthquake effects.

But, most important, this library now includes many examples of buildings around the world that were strengthened before a strong earthquake hit them. Many of these designs have been tested in actual earthquakes, and I'm happy to share their success in this new edition. A wonderful example is the huge Anheuser-Busch brewery that was strengthened by EQE just before the 1994 Northridge earthquake and is located near the epicenter of that earthquake (as discussed in the introduction to this book). I have never enjoyed a beer more than the one I had with the plant managers and engineers the day after the earthquake.

Several years ago, at a RIMS annual conference in Philadelphia, I met Andy Thompson, a young engineer who was growing the risk consulting practice in a large and prestigious international engineering and consulting firm. I began consulting with Andy, and, over the past years, we have worked together on many earthquake engineering and risk management projects throughout the world.

Andy joins me as coauthor of this new and expanded edition of *Peace of Mind in Earthquake Country*. He brings a fresh outlook, much new experience, and a broad understanding of earthquake risk management. Andy advises corporations and government organizations on how to manage their earthquake risk, from risk reduction and insurance to business continuity. He has been involved in the new and retrofit design, analysis, and assessment of numerous projects in earthquake country, including high-rises, offshore structures, and manufacturing facilities. He too uses a similar three-phase earthquake risk management program.

Together, we bring you—both homeowners and business owners—our combined experience in earthquake engineering and earthquake risk management. This new edition incorporates critical lessons learned from earthquakes since the last edition in 1991, and we structured the book around the idea of our Earthquake Risk Management Program, which is fully described in chapter 1. Most important, the book now shows what individuals and companies have done to protect themselves during the past three and a half decades, when many newly strengthened buildings, and their contents and equipment, were successfully tested by destructive earthquakes. For houses, for apartment buildings, for commercial buildings, and for industrial buildings and complexes, proactive earthquake engineering works.

INTRODUCTION

Earthquake Risk Management 101

When the first edition of this book was published in 1974, just a handful of homeowners, business owners, and government officials had taken steps to understand and manage their earthquake risk. Over the past few decades, however, innumerable residents and businesses of earthquake country have rigorously studied, reduced, and managed their earthquake risks. This new edition includes advances in knowledge made over the past three decades and is now built around our three-phase Earthquake Risk Management program.

This book functions as a step-by-step guide for achieving peace of mind in earthquake country and helps you to answer basic questions such as: Should you obtain earthquake insurance for your home or apartment? And how much should you strengthen your home or building? After going through our program for your home, you may decide that bolting your foundation and nailing down some plywood in your crawl space, together with obtaining an insurance policy, is your best solution. If you're a business owner, your solution may be more complicated, but the process of finding that solution is still straightforward.

(Figure 1)

(Figure 2)

Earthquake Risk Management for Your Home and Family

There are three phases with several steps per phase.

Phase 1: Assess Your Earthquake Risk

Phase 1 consists of assessing and studying your risk until you are satisfied that you know what to expect in a strong earthquake.

The first step is to determine the location of your property in relation to known active faults. Chapters 1 and 2 explain how earthquakes are caused, the terminology of earthquake measurement, and the appearance and hazards of active faults.

The second step is to determine the soil conditions under the property and to learn about the earthquake risks associated with soft alluvial, landfill, and sandy and water-saturated soil sites. Different soil foundations respond to a tremor with various levels of shaking intensities, and, therefore, different areas of a city such as Seattle, San Francisco, or Los Angeles face widely varying degrees of earthquake risk. Chapter 3 details how to avoid or minimize these risks.

The third step, as detailed in chapter 3, is to determine your vulnerability to landslides. Aside from active faults or unstable soil foundations near a fault, the most hazardous sites in earthquake country are landslide-prone areas, of which there are many. Strong earthquakes always trigger numerous landslides, and a property located in a known slide area could suffer landslide damage.

The fourth step, detailed in chapter 4, is to assess your exposure to other hazards. Dams present the greatest man-made danger in earthquake country; many are located in or near active fault zones, and several have collapsed during earthquakes. Levees, reservoirs, water tanks, large retaining walls, and poorly designed neighboring buildings can be equally unstable. Such structures present very real dangers in an earthquake.

The fifth step is to determine the structural strengths and weaknesses of your building. Certain architectural and structural characteristics bear directly on the amount and type of damage experienced by a building during an earthquake. A few basic and rather simple features govern the earthquake resistance of buildings, as discussed in chapters 5 and 6. And some types of buildings and structural materials either are more hazardous or are recommended for use in earthquake country. Unreinforced masonry buildings, for example, are the least resistant to earthquakes and are also the most likely to collapse. A properly braced wood-frame building, by contrast, is by far the safest small structure. Chapter 7 deals with architectural and structural features that are particularly susceptible to damage or collapse. These include houses on stilts, inadequately braced split-level homes, buildings with heavy roofs or poor foundation connections, and certain older homes. For older wood-frame houses, the two most important features are (1) anchorage of the wood sill to its concrete foundation, and (2) bracing (with plywood) of the cripple studs in the crawl space under the ground floor.

The sixth step, detailed in chapter 8, is to determine the hazards within your home—from utilities to heavy furnishings.

Phase 2: Decide How to Manage Your Earthquake Risk

Once you understand your level of exposure to earthquake risk, you must decide what to do about it.

The seventh step is to decide how and where to strengthen your building for life safety. Answers may include bolting your home to your foundation, bracing the crawl-space walls (small basement walls), protecting against other obvious collapse hazards (such as weak or "soft" stories or broad expanses of glass), and protecting against fire due to utility damage.

The eighth step is to manage your financial exposure. Should you get earthquake insurance? Should you do more strengthening to reduce damage beyond life-safety conditons? As discussed in chapter 9, it can be a good second line of defense, protecting you against financial disaster and/or unrecognized hazards such as unknown active faults or poor construction.

(Figure 1) As a homeowner, you will learn how to assess the risks to your structure. From there you can decide whether to strengthen your home—for instance by adding framing to your home's garage as shown in this image—or to buy insurance.
(Figure 2) As a business owner, you will learn whether your facility may be exposed to significant downtime in an earthquake and whether you should consider adding structural bracing as shown here.

Phase 3: Implement Your Decisions

The ninth step is to implement the strengthening work decided upon in phase 2. This may include solving safety problems in your home as well as taking steps to defend against financial losses. Throughout this book, we point out that with a few relatively simple and moderate-cost repairs and alterations, homeowners can significantly increase the earthquake resistance of their structures and substantially decrease the probability of major damage. The principles and techniques of such improvements are discussed in detail and illustrated so that, in many situations, they can be implemented by a handy do-it-yourselfer.

The tenth step is to purchase the appropriate amount of insurance, if any, as decided upon in phase 2. The eleventh step is to make a plan for what to do in the event of a major earthquake; chapter 10 provides guidance.

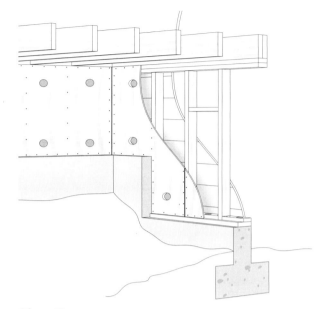

(Figure 3)

(Figure 3) One way of making your home safer is to implement strengthening work, like bracing your crawl-space walls as shown here.

EARTHQUAKE RISK-MANAGEMENT PROGRAM FOR YOUR HOME

Phase 1: Assess Your Earthquake Risk
- Step 1: Determine your property location with regard to known active faults.
- Step 2: Determine the soil conditions under the building.
- Step 3: Determine landslide potential.
- Step 4: Assess your exposure to man-made hazards.
- Step 5: Determine the structural strengths and weaknesses of your home.
- Step 6: Determine hazards in your home.

Phase 2: Decide How to Manage Your Earthquake Risk
- Step 7: Decide how and where to strengthen the existing structure for life safety.
- Step 8: Manage your financial exposure.

Phase 3: Implement Your Decisions
- Step 9: Implement strengthening work as decided in phase 2.
- Step 10: Purchase insurance, if any, as decided in phase 2.
- Step 11: Make a plan for what to do in the event of a major earthquake.

EARTHQUAKE RISK MANAGEMENT FOR YOUR HOME— A CASE STUDY

Imagine that you are in the market to buy a house and have found the one you want. It is in the Los Angeles area, was constructed in the 1940s, and is on a hill overlooking the ocean. The replacement value of the home (not the land) is $500,000. You begin by asking yourself about your earthquake exposure: Is it too close to a fault? Is it in a landslide area? What about the soil? Will my family be safe in an earthquake? Should I obtain earthquake insurance? Following the guidance of this book, you go through the process of answering these questions.

Phase 1
Using the steps outlined on page 13 and the information given later in this book, you learn that, although your property is deep in earthquake country, you are six miles away from the nearest known active fault. You also learn that although your new home is on a hill, it does not suffer a significant threat of landslide, nor is it built on liquefiable soil. You do, however, learn that, due to the date of construction, the structure was never adequately anchored to the foundation and that there are vulnerable basement ("cripple") walls in the crawl space. You also realize that that nice room on top of the garage that was added in the 1960s may pose a significant risk because the garage does not have adequate walls to support it in an earthquake. You consult a structural engineer, who confirms that indeed you must do some strengthening in the garage to alleviate this risk. You also realize that the water and gas heaters are not braced appropriately and may pose a fire risk in an earthquake. The stone chimney, too, might be a collapse hazard.

Phase 2
Through what you learn in this book, you decide that the earthquake risk to your home is manageable, and you decide to purchase the home. You should now develop a risk management plan. First, you determine what must be done to make the home safe for you and your family. You decide to anchor the house to the foundation, strengthen the basement walls, put a new steel frame in the garage, and retrofit the chimney. The cost of this work is estimated to be $35,000.

You then assess whether you should obtain earthquake insurance or implement additional strengthening that would further protect your family against financial hardship. Through consultation with your structural engineer, you learn that additional work, including strengthening the first-story walls with plywood to prevent expensive structural damage, would cost you an additional $10,000.

Your home insurance provider offers you the option of earthquake insurance. It is a member of the California Earthquake Authority (CEA), so you first consult the CEA Web site to investigate various policy options. You realize that the CEA protects against structural damage only. You look at the 15 percent deductible option ($75,000 in your case). Because of the work that you will do to keep your family safe, you are eligible for a "retrofit discount" (reducing the premium by 5 percent or about $100). Your premium is calculated by the CEA at $2,000 per year.

After reading this book and consulting your structural engineer, you realize that with the $35,000 of necessary strengthening that you plan to do, your economic loss would probably not be more than 20 percent of the replacement cost of your home (or $100,000). In this case, with your deductible, you would be liable for $75,000 for structural repair. If you implement the $10,000 of additional strengthening, your loss probably would be no greater than 10 percent of the replacement cost (or $50,000). In this case, your loss would not exceed your deductible.

You now have four options, as described in the table on page 16.

(continued)

After discussing the various options with your family, structural engineer, and insurance provider, you decide on option C because you feel that the $10,000 of additional strengthening will reduce your financial exposure and provide a safer home for your family, and because of the high deductible and low estimate of loss, insurance does not seem to be a cost-effective option.

Phase 3
You implement your decisions. You anchor your foundation, strengthen your cripple walls, and, with guidance from your structural engineer, strengthen the garage structure, repair your chimney, and provide additional strengthening to your walls. You do not purchase earthquake insurance.

OPTION	INSURANCE?	ADDITIONAL STRENGTHENING?	YOUR FINANCIAL LOSS IN A MAJOR EARTHQUAKE	YOUR TOTAL PAYOUT OVER A 25-YEAR PERIOD (ASSUMING ONE MAJOR EARTHQUAKE AND NOT ACCOUNTING FOR INFLATION)
A	No—$0/year	No—$0	$100,000 (20% of $500,000)	$100,000
B	Yes—$2,000/year	No—$0	$75,000 (limited by your deductible: 15% of $500,000)	$125,000 (25 years x $2,000 + $75,000)
C	No—$0/year	Yes—$10,000	$50,000 (10% of $500,000)	$60,000 ($10,000 + $50,000)
D	Yes—$2,000/year	Yes—$10,000	$50,000 (10% of $500,000. Your loss is limited by the actual loss, as you do not reach your deductible limit.)	$110,000 (25 years x $2,000 + $10,000 + $50,000)

Earthquake Risk Management for Your Business

Earthquake risk management for your business is an integrated approach that identifies earthquake risks, and implements a plan to accept, reduce, and/or transfer those risks. A successful plan must also be flexible, cost-effective, practical, and affordable, allowing it to be implemented and maintained through ongoing business cycles.

Both authors have used the risk management program described below for all types of companies. Many American organizations that have chosen this approach have historically concentrated on protecting their California facilities, assuming those to be their greatest exposure. But the focus is expanding to include other earthquake regions, as it should. Publicity in the past few years about high earthquake risk in both the central United States and the Pacific Northwest has prompted a number of organizations with key facilities in those areas to adopt similar programs.

Risk managers of companies with international operations also need to understand their risk should worldwide production, assembly, or delivery of their products be interrupted. Many key facilities, including many in China and all those in Taiwan and Japan, for example, are in areas of high seismicity. The earthquake risk-management program described below can be readily applied to all of these regions.

Phase 1: Assess Your Earthquake Risk

The first step is to determine the risk to your portfolio of facilities. This consists of a seismic risk survey in which financial loss estimates (often in the form of annualized loss) for various levels of earthquakes are generated. The assessment should be done for individual risk-driving facilities as well as for the company's portfolio of facilities. These assessments must utilize structural engineering and scientific knowledge, combined with the appropriate use of analytical tools and software.

The second step is to determine the risk to your facility's equipment and contents.

The third step is to understand the risk of business interruption, likely your most costly exposure. This step involves a risk analysis of business operations. Critical assets, buildings, and equipment must be identified and their associated earthquake risk highlighted. Workforce availability, supply chain, price surge, and other regional effects should also be considered.

In the aforementioned steps, potential high-risk areas are prioritized so that you can decide whether it is appropriate to continue on to phase 2 and, if so, which facilities, equipment, or operations should be included in the next analysis.

Phase 2: Decide How to Manage Your Earthquake Risk

The fourth step is to perform a detailed risk assessment of the critical risks highlighted in phase 1. You should work with an engineering company to perform this analysis of buildings and equipment with unacceptably high risks.

The fifth step is to perform cost-benefit analyses for the various options of mitigating, accepting, or transferring the risk. The earthquake-resisting capabilities of your structures and equipment should be evaluated and recommendations developed for various risk-reduction alternatives. Engineering and construction costs for implementing each option should also be estimated and offset against insurance and other risk-transfer costs. Often, annualized loss estimates can be helpful in this cost-benefit analysis of mitigation versus annual insurance premiums.

The sixth step is to develop a long-term risk-management strategy that balances risk mitigation, acceptance, and transfer through insurance or other means.

Phase 3: Implement Your Decisions

The seventh step is to strengthen your buildings, content, and equipment. This includes developing facility and equipment retrofit designs for the specific risk-reduction alternatives chosen from the options presented in phase 2.

The eighth step is to purchase the appropriate levels of earthquake insurance or utilize other risk transfer tools, as determined in phase 2. Knowing the amount of coverage that you need in your balanced program allows for more meaningful conversations with your insurance broker and insurance provider.

The ninth step is to develop contingency, emergency-response, and business-recovery plans. These plans provide a framework for company organizations and operations after an earthquake and can be kept current as facilities are modified, according to the company's long-range plans and capital budgets.

EARTHQUAKE RISK MANAGEMENT FOR YOUR BUSINESS

Phase 1: Assess Your Earthquake Risk
- Step 1: Determine portfolio facility risk.
- Step 2: Determine key risk drivers.
- Step 3: Determine extent of potential business interruption.

Phase 2: Decide How to Manage Your Earthquake Risk
- Step 4: Perform a detailed risk assessment of the critical risks highlighted in phase 1.
- Step 5: Perform cost-benefit analyses of appropriate means of managing these critical risks.
- Step 6: Develop a long-term strategy for managing your risk.

Phase 3: Implement Your Decisions
- Step 7: Strengthen building(s), contents, and equipment as determined in phase 2.
- Step 8: Purchase earthquake insurance as determined in phase 2.
- Step 9: Develop and practice emergency response and business continuity plans.

A CASE STUDY

The Anheuser-Busch Van Nuys Brewery Case Study

Anheuser-Busch operates a large brewery in Van Nuys, California, just a few miles from the epicenter of the January 17, 1994, Northridge earthquake. This facility was originally constructed in 1954 and was later expanded. In 1994, the complex served the company's markets throughout the southwest and Pacific regions, with an annual production of nearly 12 million barrels. The complex included a number of large buildings with a total replacement value of more than $1.3 billion.

Anheuser-Busch has a unique understanding of its earthquake risk. In the 1971 San Fernando earthquake, the Van Nuys brewery was damaged and beer production was interrupted for a prolonged time. During this disruption, competitors were able to make inroads into Anheuser-Busch's market share. This significant financial loss motivated Anheuser-Busch to place greater value on earthquake design in both repaired and new facilities. In fact, a new Anheuser-Busch brewery built in Fairfield, in Northern California, in the mid-1980s was designed according to earthquake standards much higher than those provided in the building code.

In the late 1980s, Anheuser-Busch initiated a comprehensive earthquake risk-reduction program for the older Van Nuys brewery to limit future damage by decreasing the vulnerability of the facility's buildings and equipment. The goal was to ensure that production following future severe earthquakes would be minimally interrupted. All of the earthquake engineering was done by EQE International.

The brewery buildings and equipment were assessed for risk, and those with unacceptable levels were seismically upgraded without affecting daily operations. Earthquake reinforcements were designed for a number of buildings and the critical equipment therein, including buildings that housed beverage production and large horizontal tanks for beer fermentation, storage, and aging. Other low-risk buildings, less important to operations and judged not to be life-safety hazards, were screened

(Figure 4) Anheuser-Busch's Van Nuys, California, brewery just after the 1994 Northridge earthquake.

(Figure 5) The brewhouse sometime in 1989 before strengthening of the building was started. The 1954 concrete frame building was found to be hazardous.

(Figure 6) The strengthened brewhouse the day after the earthquake. There was only minor damage. Note that the windows in the middle of the building are no longer there. In their place are concrete walls (shearwalls) that were added during the strengthening just before the earthquake.

out of the process, thereby ensuring the most efficient use of limited resources. The total cost of the strengthening program, 90 percent of which was for construction, was about $11 million, less than 1 percent of the total facility replacement cost.

The 1994 Northridge earthquake produced very strong ground motion, causing partial collapses and extensive damage to many buildings in the brewery's immediate vicinity. Post-earthquake surveys conducted by Anheuser-Busch's engineering consultants showed that none of the retrofitted structures sustained significant damage, nor did equipment essential to brewery operations. Additionally, no major employee injuries were associated with the earthquake. Other on-site buildings and equipment that had not been strengthened in the 1980s did sustain damage, requiring about $17 million in repairs. The brewery returned to nearly full operation in seven days following minor cleanup and repairs and restoration of the off-site water supply. Anheuser-Busch lost none of its pre-earthquake market share, which had been the overriding goal of the earthquake risk-reduction program.

Anheuser-Busch estimated that its facility would have suffered direct property losses of about $350 million had it done no earthquake strengthening. This averted damage was worth more than thirty times the actual cost of the brewery's loss-control program. However, when business interruption and other indirect losses such as market share are considered, total losses could have exceeded $1 billion had no strengthening efforts been implemented. Clearly, Anheuser-Busch's earthquake riskmanagement program paid for itself many times over. This case study shows how proactive earthquake risk-management can protect both corporate balance sheets and employees.

(Figure 4)

(Figure 5)

(Figure 6)

1906 San Francisco earthquake

CHAPTER 1

A Primer on Earthquakes:
How Earthquakes Are Caused and Measured

Until recently, the cause of earthquakes remained a profound mystery. The first milestone on our path toward knowledge dates to November 1, 1755, when a great earthquake struck near Lisbon, Portugal. The city fell in ruins, and up to sixty thousand people perished. After the quake, Portuguese priests were asked to perform a survey of the damage and document their observations. Their records represent the first systematic attempt to investigate an earthquake and its effects.

Before that time, tremors in the earth were generally relegated to that category of natural misfortunes that insurance companies still call "acts of God." Aristotle proposed that the frequent quakes that shook the ancient Greek temples and cities were caused by gales that became trapped in giant subterranean caves. The Greek philosopher Anaxagoras reasoned that earthquake motion occurred when large sections of the earth cracked and tumbled into the hollow terrestrial core. The Roman scholar Pliny said that earthquakes were simply Mother Nature's method of protesting the wickedness of men who mined gold, silver, and iron ores. Other ancient philosophers placed the blame on an ill-tempered Poseidon, god of the sea and the watery element.

The northern neighbors of these philosophers—the tribes of what is now Bulgaria—believed that earthquakes struck when an enormous water buffalo, which carried the world on its back, readjusted its burden to ease its task. Other peoples throughout the centuries have also assigned the phenomenon to monstrous mythic animals—giant hogs, catfish, tortoises, spiders, frogs, whales, serpents—whose occasional restlessness caused the world to tremble. As charming as these ancient notions are, scientists of later centuries had little better to offer. An Italian scholar in the sixteenth century suggested that the best form of earthquake protection was to place statues of Mercury and Saturn on a building. Even in the scientific age of the eighteenth century, most observers were, like their ancient predecessors, satisfied that earthquakes were not explainable except as a capricious force of nature or the stern work of an offended god. In Puritan New England, sermons with such titles as "Earthquakes: a Token of the Righteous Anger of God" and "The Lord's Voice in the Earthquake Crieth to Careless and Secure Sinners" were said to be validated by the fact that the 1755 earthquake near Catholic Lisbon was far more destructive than the Cape Ann earthquake near Protestant Boston of that same year (even though the Boston quake was a great deal smaller in magnitude). People's task was merely to wait and pray (and try to be good).

The Source of Earthquakes—The Theory of Continental Drift and Plate Tectonics

Even today, the causes of earthquakes are not completely understood. But there is now sufficient scientific evidence to conclude that the tremors are the effect of a rebalancing of forces arising from the collision of continuously moving plates on the earth's surface. This idea is based on the theory of plate tectonics, developed in the 1960s, which incorporates older theories of continental drift and the concept of seafloor spreading.

(Figure 1) California's San Andreas fault is one segment of the line of intersection between the North Pacific and the North American tectonic plates. Both plates are moving slowly north and west at different rates, producing the frictions and temporary locks along the fault that are released in the sudden shifts of earthquakes, the surface distortions of the broad fault zone, and the gradual growth of the coastal ranges. The Sierra Nevada ranges were formed when the two plates collided directly, and the thinner Pacific plate was forced downward, buckling the continental plate, lifting the mountain range and forming the westernmost portion of California with an accretion of materials from the oceanic plates. The descending remnant of the ancient plate is representative of a blind thrust fault like the ones that caused the 1994 Northridge, Los Angeles, quake.

(Figure 2) The same geological changes that cause earthquakes to occur are also a source of much natural beauty. This is illustrated at 1000 Island Lake on the John Muir Trail in the California Sierras.

(Figure 3) The dark areas on this map indicate the distribution and relative density of earthquakes recorded throughout the world. These belts of seismic activity mark with dramatic clarity the turbulent boundaries of the drifting, colliding tectonic plates that form the earth's crust. The mid-oceanic lines of activity represent the towering mountain ranges and deep rift valleys where the younger tectonic plates are renewed and pushed outward, altering the sea floors a few inches every year. About 80 percent of the planet's earthquakes occur along the Circum-Pacific seismic belt, which loops completely around the Pacific Basin. The Alpide belt, which extends from Java through the Himalayas and into the Mediterranean is responsible for about 17 percent of the world's seismic activity. The remaining 3 percent of all earthquakes strike along the Mid-Atlantic Ridge and in scattered pockets of seismic activity throughout the world.

The theory of plate tectonics states that the outermost part of the earth is made up of two layers: the lithosphere and asthenosphere. The lithosphere "flows" atop the relatively fluid (geologically speaking, that is; it is actually pretty solid stuff) asthenosphere, and is broken up into several major, and many minor, tectonic plates, which are up to approximately sixty miles thick. As these plates collide and move against one another, mountains form, volcanoes erupt, and earthquakes relieve the built-up frictional forces that resist their movement. The plates move laterally at typical speeds of a few inches per year, the usual rate at which fingernails grow.

The newest and thinnest of these tectonic plates are the ocean floors, which are still being formed from molten materials flowing from the earth's interior. This flow emerges in deep rift valleys that form the inner boundaries of the suboceanic tectonic plates and divide vast, continuous mountain ranges that traverse the length of all the ocean basins. Molten materials from the earth's interior well up through the rift valleys and solidify to build the edges of the oceanic plates. These young oceanic plates are then pushed slowly but steadily away from the rift valleys, pressing their outer edges against the established and heavier plates that make up the continental land masses. It has been demonstrated, for example, that the Atlantic Ocean is spreading from the Mid-Atlantic Ridge at about one inch a year, so that within an average person's lifetime, the continents of Europe and North America move about six feet farther apart.

As the oceanic plates meet the continental plates, tremendous pressures buckle the earth's surface (creating mountain ranges); plunge the thinner, weaker oceanic plates into deep-sea

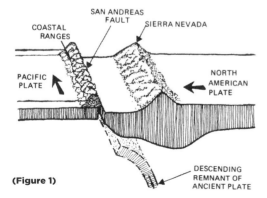

(Figure 1)

(Figure 2)

trenches beyond the continental shelves; and trigger volcanoes and earthquakes. Along the western coast of South America, for example, the thinner oceanic plate is forced downward by the thicker and heavier continent. As it is propelled below the continental plate and melts into the earth's core, the Andes mountains are pushed continually upward. At the same time, friction causes a temporary lock between the two plates. The inevitable and frequent failures of this fragile bond cause the deep, powerful earthquakes typical of Chile and Peru. A similar type of collision can occur between two thick continental plates as well. For example, the subcontinent of India is a separate plate that is moving northward against the Asian mass. The soaring Himalayas, as well as such destructive tremors as the 2001 Bhuj (India), 2005 Pakistan, and 2008 Wenchuan (China) earthquakes, are the result.

Some of the largest faults—breaks in the rock of the earth's upper crust—are formed in the region of the line of collision between tectonic plates. The San Andreas fault system of California is the result of the ancient and continuing collision of the North Pacific plate and the North American continent. Many millions of years ago, the more massive westward-moving continental plate overrode the opposing Pacific plate, driving the latter downward into the earth's crust, pushing up the Sierras, and causing the violent blowouts of such volcanoes as St. Helens, Shasta, Lassen, Rainier, and Hood. At the same time, some of the plunging Pacific plate was scraped off against the continent at the San Andreas fault zone, so that the coastal surface of western North America grew outward by about one hundred miles in a very gradual accretion of new materials, forming much of California and its coastal ranges. Thus, the southwestern third of the state west of the San Andreas fault is made up of relatively new geologic materials riding the Pacific tectonic plate, while the remainder of the state forms the western edge of the North American plate.

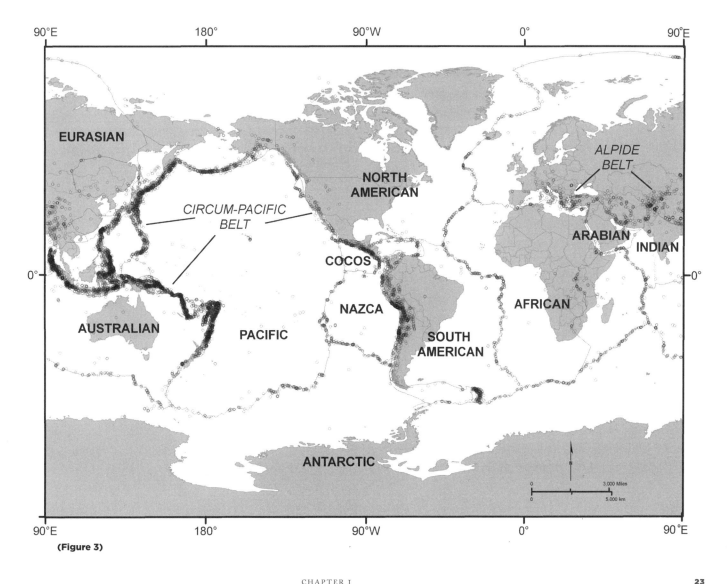

(Figure 3)

Today, these two plates have changed directions, so that they are essentially sliding past each other along the San Andreas fault. The great Pacific plate carries the ocean floor, a part of California, and all of the Baja peninsula northwestward in relation to North America, while the North American plate, pushed by the seafloor spreading at the Mid-Atlantic Ridge, moves west at a slower rate. The two plates finally collide directly in the far north, along the Aleutian archipelago, where the Pacific plate is driven downward. It is estimated that at the present rate of movement, the Los Angeles area, riding the Pacific plate, will draw abreast of the San Francisco Bay Area in about ten million years (no doubt to each other's dismay).

The Mechanism of Earthquakes—
The Theory of Elastic Rebound

The edges of the plates have a certain amount of elasticity and tend to hold their positions along the fault. Portions of the fault frequently remain locked in this way, under tremendous stress, for several years or even centuries. Finally, when the accumulated sliding force exceeds the frictional force that binds portions of the plates and prevents their natural movement, the distorted rock along the two sides of the fault suddenly slip past each other in an explosion of movement that allows a new position of equilibrium.

This slippage, termed elastic rebound, produces powerful vibrations, sometimes ruptures the earth's surface and may shift the positions of the two sides of the fault by several feet both horizontally and vertically. Earthquakes are the result of these violent adjustments of a temporarily locked fault.

Two types of earthquakes are associated with different types of plate collisions. Shallow-focus earthquakes, with an average depth of three to ten miles below the earth's surface, result from the slippage of primarily laterally moving plates and are typical of California and most of the seismic regions of the American West. Both the 1989 Loma Prieta and 1994 Northridge foci were approximately eleven miles deep. Deep-focus earthquakes usually occur where plates collide directly and one is forced below the other. For example, the great Chile quake of 1960 and the 2001 Nisqually, Seattle, earthquake both occurred at a depth of about thirty miles. The 2004 Indonesia earthquake (which caused the devastating Indian Ocean tsunamis) originated at a depth of about twenty miles. Many other earthquakes in subduction areas occur at depths greater than seventy-five miles.

An earthquake's destructiveness is closely related to its depth: The shock waves of deeper earthquakes generally dissipate as they rise to the surface and are therefore less damaging. On the other hand, deep-focus tremors usually affect a much wider area. Shallow-focus earthquakes are felt over a smaller area but are sharper and usually more destructive. For example, earthquakes in the Puget Sound area of Washington have depths typically three to five times greater than those of equally large earthquakes along the San Andreas fault, and historically, these shocks have been, so far, considerably less destructive than those in California.

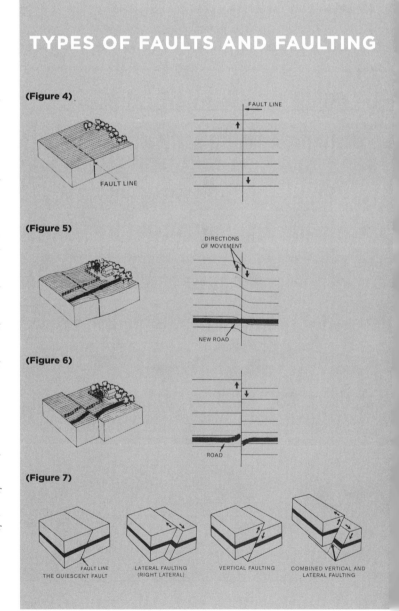

TYPES OF FAULTS AND FAULTING

(Figure 4)

(Figure 5)

(Figure 6)

(Figure 7)

(Figure 4)
The quiescent fault
A few faults may move relatively freely and very slowly along the plane of the drifting tectonic plates. This movement is termed fault creep.

The Calaveras fault and a portion of the San Andreas fault near Hollister in Northern California have moved in this way. Other faults become locked with the friction of the colliding plates and move only when the rocky layers of the plates become strained beyond endurance, then slip apart with the violence of an earthquake.

(Figure 5)
The strained fault before an earthquake
The gradual movement of the tectonic plates has created strain (or stored energy) in the rock of the fault where the two plates meet. The frictional force of the collision locks the two sides of the fault and prevents any movement. The limited elasticity of the rock allows the strains of this locked fault to accumulate for decades. Finally, the rocks give way, allowing the two sides of the fault to realign and causing the upheaval of an earthquake and surface and below-surface displacements.

(Figure 6)
The adjusted fault after an earthquake
The fault has moved into a new, unstrained position, causing surface displacements that have destroyed the continuity of the highway and fence and producing intense shock waves during the quake that have demolished poorly engineered buildings near the fault zone.

(Figure 7)
The direction of faulting
Faults typically move either laterally, vertically (thrust and/or graben faulting), or in a combination of vertical and lateral shifts. The San Fernando and White Wolf faults in Southern California fit this latter category of movement, which is quite common. The faults of the San Andreas fault system move primarily laterally. The Wasatch fault in Utah and the Kern River and Pleito faults in Southern California are vertically-moving faults.

DON'T BE FOOLED BY THE 2001 NISQUALLY EARTHQUAKE

Was the magnitude 6.8 Nisqually earthquake of 2001 the largest that the Puget Sound area can expect? No, not even close. Earthquakes of magnitude 9 or greater are possible here, and buildings in the region, even the newest ones, simply were not (and are not) designed for this level of earthquake.

The Nisqually earthquake was a deep interplate earthquake, originating a full thirty miles beneath the surface. By the time the earthquake waves reached the surface, their damaging effects had been greatly diminished. These types of deep earthquakes occur in the region approximately every thirty to fifty years, most recently in 1949, 1965, and 2001. There is an 85 percent chance of another earthquake of this caliber striking the Puget Sound region in the next fifty years.

Earthquakes resulting from shallower faults in the region, like the Seattle fault, would have a much greater impact than the Nisqually one did. A recent study of possible effects of a Seattle fault zone rupture predicted sixteen hundred deaths and $33 billion in damage, and much higher numbers are possible. There is a 15 percent chance that an earthquake of this type could occur in the next fifty years.

Potentially causing even greater damage would be a rupture along the Cascadia subduction zone, a fault spanning from Vancouver Island to Northern California. Caused by the North American plate colliding with and sliding over the Juan de Fuca plate, this fault could produce earthquakes of magnitude 9 and greater. Again, there is as much as a 15 percent chance of such an earthquake occurring over the next fifty years. The Pacific Northwest and British Columbia have not experienced this type of an earthquake, or disaster, in modern history. Such a disaster could be as destructive to the economy of the area as the 2005 Hurricane Katrina was to the economy of New Orleans.

Seattle, Washington

Faults, Fault Zones, Faulting, and Creep

Fault and *fault system* are the terms used to describe not only the demarcation of opposing tectonic plates, such as the major trace of the San Andreas, but also the related web of numerous crustal breaks that result from the stresses of plate collisions and the widespread subterranean damage of frequent earthquakes along the collision course. California, for example, is made up of a network of blocks that move in different directions as a result of the collision of the North American and Pacific plates and the activity of the San Andreas fault. Each block is separated from the others by a large fault, and most of these faults are active and therefore capable of abrupt slippage. In addition, throughout each crustal block are lines of fracture that form lesser faults, also under pressure and capable of slippage and eruption.

Thus, the plane along which slippage and tremors occur may be a recently active major fault or a newly created line of fracture in the weakened rock of older and presumably inactive ("healed" or unstrained) fault traces. For example, few geologists suspected any earthquake potential along the minor fault in the San Gabriel Mountains behind Los Angeles until it ruptured on February 9, 1971, thrusting some of the mountains 6 feet higher, killing sixty-four persons, and demolishing hundreds of buildings in the San Fernando Valley. Twenty-three years later, on January 17, 1994, the region was rocked again by a barely suspected fault, this time a thrust fault along the northern fringes of the San Fernando Valley in Northridge, Los Angeles County; the quake killed fifty-seven people. Similarly, the 1983 Coalinga earthquake in Central California and the Whittier, Los Angeles, tremor of 1987 occurred on so-called blind thrust faults that are hidden underground.

Faults are generally narrow, measuring only a few inches to several feet wide. However, a major fault system, such as the San Andreas, includes not only the most recent active break but also a broad fault zone of shattered rock and traces of previous breaks and surface ruptures. The San Andreas fault zone, usually many hundreds of feet wide, reaches more than 1.5 miles in width at numerous locations along its 650-mile length.

When the stresses of a fault are released in an earthquake, the highest-intensity shock waves and vibrations are usually felt near the fault line and nearest the point of slippage. However, much of the fault's length may be affected, so that destructive vibrations can occur for many miles on either side of the earthquake center. In addition, the shock waves disperse from the fault like the rings

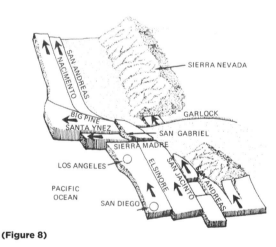

(Figure 8)

(Figure 9)

(Figure 8) This drawing of the San Andreas fault system in Central and Southern California illustrates a few of the numerous small crustal blocks and fault divisions created by the stresses of the tectonic plates collision and the frequent upheavals along the San Andreas. These lesser faults generally parallel the lateral northwesterly movement of the San Andreas and the continental and Pacific plates. However, this parallel movement is interrupted along the intersection with the Garlock fault, where the southern group of blocks encounters the deep crustal roots of the Sierra Nevada and is deflected to the west.

(Figure 9) The San Andreas fault through California is often an easily distinguished linear scar across the landscape. The fault is not so easily distinguished once development and building occur in the fault zone.

produced by a pebble dropped into still water, so that significant shocks and damage can affect areas for many miles on either side of the fault. The intensity of shaking diminishes with distance, of course, but softer, vibration-prone soil many miles from the fault can produce more damage than strong rock formations only a few hundred feet from the source of the quake can.

The energy waves generated by an earthquake are modified by reflection and refraction through the different layers of rock and soil and the various geologic features (mountains, hills, plains) that make up the surface of the earth surrounding the fault. The combination of "pushes and pulls," "ups and downs," and reflections and refractions of these waves by the ground itself creates the chaotic and violent surface motions of a strong earthquake.

During a strong tremor, the ground surface along a fault may rupture and shift laterally or vertically by several feet. The 1906 San Francisco quake on the San Andreas fault resulted in lateral fault shifts of as much as 15 feet. Other earthquakes, such as the tremendously large shock in Owens Valley, California, in 1872 and the more moderate Hebgen Lake, Montana, earthquake in 1959, caused vertical displacements of 20 feet or more at the surface. This displacement of ground along a fault during a quake is termed *faulting* and results in the abnormal topographical formations of scarred, crumpled, and upthrusted rock and soil that visibly delineate a major and recently active fault. Faulting also destroys or severely damages structures. A small earthquake, by contrast, indicates a localized readjustment of fault strain, and usually such shocks are not accompanied by either the surface displacements of faulting or serious damage to buildings.

Another type of fault movement, fault creep, occurs when the two sides of a section of a fault do not lock completely but move past each other at an infinitesimally slow and gradual rate (mere fractions of an inch per year). Occasionally, the creep may halt entirely for a time, or it may drastically increase its speed, with spurts of discernible movement over a few days. Whatever the speed of movement, the gradual surface displacements of fault creep can damage buildings in the immediate vicinity of the fault. Fault creep may also be accompanied by barely perceptible micro-tremors.

Some scientists believe that the continuous adjustment of fault stresses through creep reduces the maximum magnitude of a future earthquake along a fault's creeping segment. Most, however, agree that creeping faults, such as the Hayward fault in the San Francisco East Bay and the Calaveras and San Andreas faults near Hollister, are highly susceptible to large surface displacements in the event of a major quake.

(Figure 10)

(Figure 11)

(Figures 10 and 11) A fence in Olema, north of San Francisco, was split by the 1906 rupture along the San Andreas fault. The foreground moved about 15 feet to the left (north). The area is now part of the Earthquake Trail at the Point Reyes National Seashore. Fault creep cannot be seen when comparing an image (figure 10) from the 1991 publication of this book with another of present day (figure 11).

(Figure 12)

(Figure 13)

(Figure 14)

(Figure 15)

(Figure 12) The 1987 Bay of Plenty, New Zealand, earthquake, with a magnitude of 6.2, was accompanied by about 6 miles of surface faulting. The maximum vertical uplift was about 6 feet. Well designed single-story wood-frame houses within a few feet of the fault were usually undamaged. Two-story houses and poorly reinforced masonry buildings were not as fortunate.

(Figure 13) Evidence of fault damage can easily be erased. The same site about ten years later.

(Figure 14) Faulting along the Landers fault in Southern California in the M7.6 1992 earthquake.

(Figure 15) One of the best-known major buildings on a lateral fault is the University of California, Berkeley, football stadium. Along with several other structures near the campus, it straddles this creeping fault; slowly widening cracks indicate that the structure is gradually being split apart by creep. This figure shows visible evidence (the vertical gap to the left of coauthor Andy Thompson) of the fault's creep movement at the upper south end of the stadium seats in what is the Stanford section during the annual Big Game.

(Figures 16 and 17) The 1987 New Zealand earthquake caused this faulting. These are before (figure 16) and after (figure 17) photographs of a residential driveway.

(Figures 18 and 19) Horizontal and vertical faulting: Figure 18 shows faulting along the well-known North Anatolia fault (often compared by scientists and engineers to the San Andreas) in the M7.6 Turkey earthquake of 1999. Note the offset center line of the road. There was hardly any vertical movement of the fault in this area. Figure 19 shows the same fault caused a 7-foot vertical fault scarp at a nearby location.

(Figure 16)

(Figure 17)

(Figure 18)

(Figure 19)

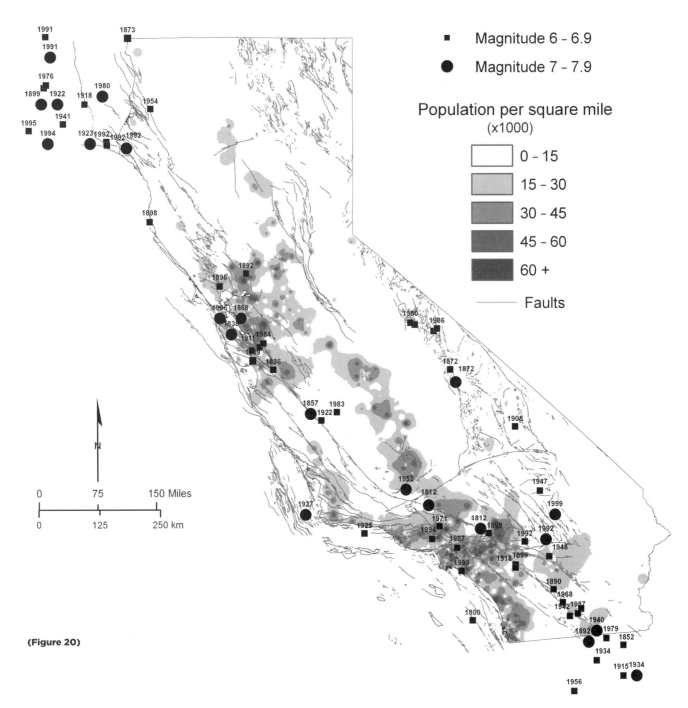

(Figure 20) This map of the major and lesser faults and earthquakes of California clearly shows that few populated areas in the state are unaffected by earthquake danger.

(Figures 21 and 22) This large single-story building was seriously damaged in the strong M6.4 Coalinga, California, earthquake of 1983. The weakened brick structure was then destroyed by aftershocks, as is shown in the second photograph.

(Figures 23 and 24) This two-story unreinforced masonry and wood house was also seriously damaged by the 1983 Coalinga quake. The house collapsed two days later from aftershocks.

Foreshocks and Aftershocks

A large earthquake rarely consists of a single shock. A series of foreshocks sometimes occurs before the main shock, and aftershocks may occur following the main shock over the span of several months and can still be considered part of one earthquake.

Our present technology cannot adequately distinguish foreshocks from the numerous other small earthquakes that occur frequently in seismically active regions. Thus, foreshocks cannot yet serve as warnings of impending danger. However, seismologists might someday detect certain characteristics that identify foreshocks and use them as early warning signs of the size and location of larger future quakes.

Aftershocks are caused by the continuing readjustment of stresses at different locations along a ruptured fault and its subterranean fault plane after the main shock. Because of varied geology along the fault plane, all of the accumulated energy is not released at once by a quake, and the process of localized readjustment continues indefinitely.

The San Fernando area, for instance, was disturbed by thirty-three large aftershocks in the first hour after the severe 1971 earthquake there. Even a year after the main shock, on February 9, 1972, a minor aftershock struck. The October 17, 1989, Loma Prieta earthquake had a main shock of magnitude 6.9. In the next twenty days, there were 4,760 aftershocks, two of them damaging, twenty strong, and sixty-five perceptible to people. The 1994 Northridge earthquake was followed by more than ten thousand aftershocks, two of which were greater than magnitude 6.0.

Some large aftershocks can cause a great deal of building damage. For example, a month after the major Kern County earthquake in 1952, an aftershock centered near Bakersfield caused more damage to the shaken and weakened city than the initial quake had, because it struck much nearer to the city.

(Figure 21)

(Figure 22)

(Figure 23)

(Figure 24)

(Figure 25)

(Figure 26)

The Measurement of Earthquakes

Epicenters and Hypocenters

News media reports on an earthquake inevitably give detailed accounts about the epicenter of the shock, and the public usually assumes that the epicenter is the area most seriously shaken by the earthquake. The epicenter is very definitely related to the strength of ground shaking (known as intensity). However, other factors, such as depth, geologic foundations, length of the fault rupture, and the extent of faulting, are much more important to intensity.

The location, deep in the crust of the earth, where a fault slippage begins is known as the hypocenter, or focus, of the earthquake. The epicenter is the projection of the hypocenter on the ground surface and is always the point on the surface closest to the initial slippage. However, the epicenter should not be confused with the point in the affected area that experiences the strongest or longest shaking; depending on the angle of the fault through the bedrock, the surface point nearest the quake's focus may be several miles away from the fault. And the vicinity of the fault, not the epicenter, is almost always the center of greatest quake intensity and, therefore, has the greatest potential for damage.

It is important to also consider the response of an entire fault line. For example, in the 1906 earthquake, the epicenter was located just southwest of the Golden Gate, but the intensity of the shock was equally strong along the fault in San Jose (more than 45 miles from the epicenter). Faulting and intense shock waves also hit as far south as Hollister (90 miles from the epicenter). If you wish to estimate the future earthquake hazard of a specific place, you will want to consider its distance from an active fault as a more important criterion than the location of epicenters along that fault.

(Figure 25) This figure shows the relationship between the hypocenter and the epicenter of an earthquake occurring on a lateral fault, such as the San Andreas fault in California.
(Figure 26) From the standpoint of ground shaking intensity and building damage, the location of the epicenter of an earthquake can be very misleading. The epicenter of the famous 1906 quake was thought to be almost directly on the San Andreas fault near the village of Olema although it is now known to be closer to the Golden Gate. However, the fault ruptured and caused large surface displacements and shock waves that leveled much of Hollister, nearly 100 miles south along the fault from the epicenter.

The Magnitude Scale

The first question asked after an earthquake is "How big was it?" The answer is not as straightforward as the question. Earthquakes are usually measured by two very different scales—the Moment Magnitude scale and the Modified Mercalli Intensity (MMI) scale—and both are often confused by the public.

Earthquake magnitude, a measure of the total energy released by a quake, was originally defined in 1935 by professors Beno Francis Gutenberg and Charles E. Richter of the California Institute of Technology in Pasadena, following up on work by Japanese researchers. Originally, the magnitude scale was called the Gutenberg-Richter scale, and later it came to be known as the Richter scale. Because there was no way to directly measure the released energy of an earthquake, Gutenberg and Richter based their scale on the alternations of a sensitive movement-measuring instrument—a seismograph—hypothetically located 62 miles (100 km) from the center of surface energy released by the shock (the epicenter). Since in reality the distance from an epicenter to a seismic recording station is never exactly 62 miles, mathematical tables are used to convert seismograph records into standard whole numbers and decimals on a scale of 1 to 9. Advances in the seismological sciences since the introduction of the Richter scale now permit the magnitude of an earthquake anywhere in the world to be reported within minutes of its occurrence. The Richter magnitude scale (or local magnitude scale, ML, as it is now called) works reasonably well for small to moderate-size earthquakes, but poorly for very large ones.

The magnitude scale most useful to professionals today is the Moment Magnitude scale, or Mw, which comes closest to measuring the true size of an earthquake, particularly a very large one, such as the magnitude 9.1 (sometimes abbreviated as "M9.1") Indonesia earthquake in 2004, which caused a series of devastating tsunamis in the Indian Ocean. This scale relates magnitude to the ruptured area of the fault and the amount of slip movement occurring on the fault. Gradually, the media and the public are beginning to use "magnitude," a more general term that usually refers to Moment Magnitude, as opposed to stating a quake's outdated Richter magnitude. It is therefore generally incorrect to use the term "Richter Scale."

The magnitude scale is used to compare the sizes of earthquakes, but the figures can be misleading without an understanding of the scale's mathematical basis. The magnitude scale is logarithmic, with each whole number representing a magnitude of energy release that is approximately 31.5 ($10^{1.5}$) times the lower number. This means that there is 31.5 times more shaking energy in an earthquake of magnitude 6 than in one of magnitude 5, and roughly 1,000 times more in a magnitude 7 earthquake.

People occasionally believe that the occurrence of many small quakes in one area depletes fault stresses and therefore reduces the possibility of a large shock. This is a myth, for with the magnitude logarithmic scale, it would take over fifty shocks the size of the Northridge earthquake (magnitude 6.7) or more than thirty Loma Prieta earthquakes (magnitude 6.9) to equal the energy released by the single great San Francisco earthquake of 1906 (magnitude 7.9). On the same basis, it would require over five thousand Northridge earthquakes to equal the energy of the 1964 Alaska earthquake (magnitude 9.2). Clearly, we are better off with one large quake than many smaller but still destructive ones.

The Modified Mercalli Intensity Scale

Sometimes it is more useful to describe the effects of an earthquake rather than the energy release of an earthquake. Intensity maps can graphically illustrate the varied intensity of ground shaking in different earthquakes, something not provided by magnitude scales. There are several intensity scales, all based on reports of ground and building damage and on interviews with people in various locations in the earthquake-affected areas. These scales were developed before earthquake-recording instruments were available as a means to evaluate the relative size of an earthquake. Various categories of earthquake damage, ground effects, and personal sensations, emotions, and observations are defined and assigned numerical designations.

The Modified Mercalli Intensity (MMI) scale is the one most commonly used in the United States. The MMI scale is denoted with Roman numerals from I to XII, with each number corresponding to descriptions of earthquake damage and other effects. Because damage and ground effects are influenced by numerous factors—such as distance from the causative fault, the type of soil beneath the observer, the type of building, the accuracy of personal observations, and so forth—reported intensities vary considerably from site to site, with large differences sometimes occurring at locations only a few feet apart.

Earthquake intensities observed at various locations are plotted on an intensity, or isoseismal, map. The intensity maps illustrated below for the 1906 San Francisco and 1989 Loma Prieta earthquakes are typical of these types of maps. The divisions (isoseismal lines) between the intensity zones form an oval pattern about the focal area of the quake. In comparison, the intensity map of the 1906 San Francisco quake covers a much larger area and shows elongated isoseismal lines, reflecting the greater magnitude of the earthquake, the effect of the shock waves that emerged along much of the ruptured length of the San Andreas fault, and the considerable amount of faulting. The isoseismal lines between intensity zones are always, at best, a very rough approximation of the boundaries of the various intensities; it would be impossible to include all of the reported variations in damage, observations, and sensations for a particular quake.

Intensity maps like these are usually published by the U.S. Geological Survey (USGS) soon after a major earthquake strikes a populated area. In addition to these large-scale maps of whole affected areas, very detailed intensity maps of individual cities, and even subdivisions, may be plotted for a quake. The latter can be very useful to an urban property owner in determining the range of shock wave intensities and damage that can be expected during the next comparably sized earthquake.

Because the MMI scale and the Moment Magnitude scale measure entirely different parameters, the two cannot be readily compared. The magnitude scale records physical energy with instruments and therefore gives no consideration to the important factor of geologic conditions. The MMI scale, on the other hand, is based solely on observations. A simplified summary of the MMI scale is opposite.

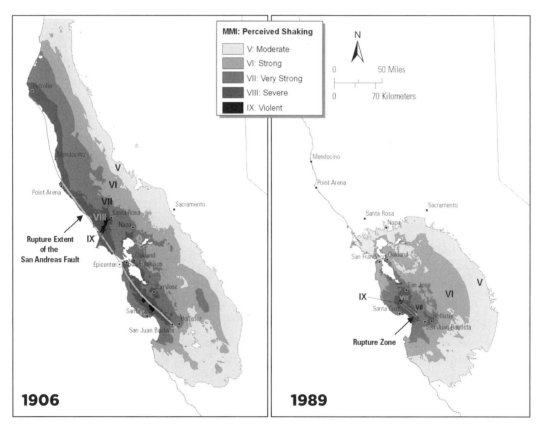

(Figure 27)

The Modified Mercalli Intensity (MMI) Scale (abridged)

I. Not felt by most people; only instruments detect the earthquake.
II. People lying down might feel the earthquake.
III. People on upper floors of building will feel it but may not know it's an earthquake. Hanging objects swing.
IV. People indoors will probably feel it, but those outside may not. Houses may creak.
V. Nearly everyone feels it. Sleepers are awakened. Doors swing; pictures move; things tip over.
VI. Everyone feels the earthquake. It is hard to walk. Windows and dishes broken. Books fall from shelf.
VII. It is hard to stand. Plaster, bricks, and tiles fall from buildings. Small landslides.
VIII. People will not be able to drive cars. Poorly built buildings may collapse; chimneys may fall.
IX. Most foundations are damaged. Masonry heavily damaged. Pipes are broken. The ground cracks.
X. Most buildings are destroyed. Water is thrown out of rivers and lakes. Large landslides.
XI. Rails are bent. Bridges and underground pipelines are unusable.
XII. Most things are leveled. Large objects may be thrown into the air. Large rock masses are displaced.

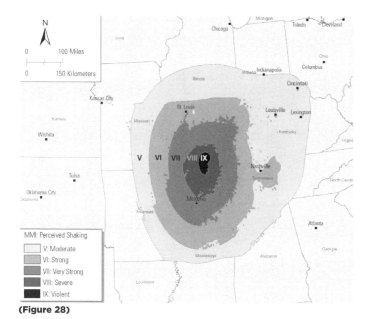

(Figure 28)

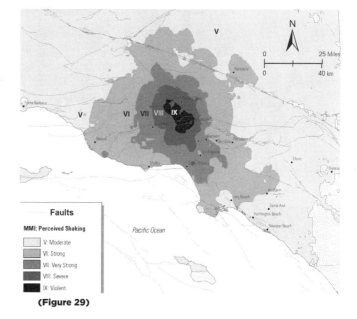

(Figure 29)

(Figure 27) After a tremor, earthquake scientists and engineers gather the public reports of structural damage, ground effects, interior damage, and personal observations and sensations and then plot this data on intensity, or isoseismal, maps such as these: from the April 18, 1906, San Francisco earthquake (left), and the October 17, 1989, Loma Prieta earthquake (right). This clearly illustrates the difference in energy release between the two earthquakes.

(Figure 28) Shown here is the MMI map for the New Madrid earthquake of 1812. The area of shaking is ten times larger than that of the 1906 San Francisco earthquake. Although this is the largest earthquake to have hit the United States since settlement by Europeans, buildings in the area are not designed for earthquakes to the same requirements as in California. This issue is discussed in greater detail later in the book.

(Figure 29) Shown here is the MMI map for the 1994 Northridge earthquake. Although this caused significant human and financial losses, it was not nearly as strong as the 1906 San Francisco nor the 1812 New Madrid earthquakes.

(Figure 30)

(Figure 31)

(Figure 32)

(Figures 30-32) The power of earthquakes and the severity of ground shaking were illustrated graphically by the following examples of trains being overturned. The first example (figure 30) is from the San Francisco earthquake of 1906. The second example (figure 31) shows a derailed Bullet Train in the 2004 Niigata Ken Chuetsu, Japan, quake west of Tokyo. The third example (figure 32) shows a stationary 67-ton diesel train engine that was overturned by the 1987 Bay of Plenty, New Zealand, earthquake.

EXPERIENCING A STRONG QUAKE—THE GREAT SAN FRANCISCO EARTHQUAKE OF 1906

We have yet to experience a major earthquake in the mainland United States in the past 100+ years (the 1989 and 1994 California earthquakes, for example, were not major earthquakes). We must therefore rely on accounts from the 1906 San Francisco earthquake (as well as accounts from the other devastating earthquakes that occur regularly thoughout the world). Following are some dramatic accounts from the San Francisco earthquake of 1906.

Jesse B. Cook:
". . . The whole street was undulating. It was as if the waves of the ocean were coming toward me, and billowing as they came. . . . Davis Street split right in front of me. A gaping trench that I think was about six feet deep and half full of water had suddenly yawned between me and the east side of the street. It seemed to extend for blocks. . . . I simply took it at a running jump, and sprang up on the sidewalk. . . . As I stepped down from the sidewalk, I looked for the trench I had been obliged to jump over. It was gone; closed up again; and I had never noticed it closing."

Ernest H. Adams:
". . . I was thrown out of bed and in a twinkling of an eye the side of our house was dashed to the ground. How we got into the street, I will never be able to tell, as I fell and crawled down the stairs amid flying glass and timber and plaster."

Peter Bacigalupi:
". . . I was awakened from a sound slumber by a terrific trembling, which acted in the same manner as would a bucking bronco. I sat up in bed with a start. My bed was going up and down in all four directions at once, while all about me I heard screams, wails, and crashing of breaking chinaware and knickknacks. I was very quietly watching the clock on the mantel, which was doing a fancy stunt, while the ornaments in the parlor could be heard crashing to the floor. A great portion of plaster right over the head of my bed fell all around me and caused a cloud of dust, which was very hard to breathe through."

Fred J. Hewitt:
"It is impossible to judge the length of that shock. To me it seemed an eternity. I was thrown prone on my back, and the pavement pulsated like a living thing. Around me the huge buildings, looming up more terrible because of the queer dance they were performing, wobbled and veered. Crash followed crash and resounded on all sides. Screeches rent the air as terrified humanity streamed out into the open in an agony of despair."

San Francisco earthquake, 1906

1989 Loma Prieta earthquake, Santa Cruz Mountains

1999 Mormara, Turkey, earthquake

CHAPTER 2

The Hazards of Faults and Faulting

Buildings in fault zones are exposed to very high earthquake risk. No measures—whether the most earthquake-resistant bracing and building materials or the latest and soundest principles of reinforcement—can guarantee that a property astride a fault will survive a moderate quake without serious damage. In a study of damaged and undamaged houses on two similar streets in Sylmar after the San Fernando quake in 1971, 80 percent of the houses in the fault zone suffered moderate or worse damage, whereas only 30 percent of the buildings immediately beyond the zone suffered such damage.

The greatest hazard to structures in fault zones is a ground-surface rupture, and no reasonable building normally can withstand such faulting beneath it. A ground shift of only a few inches (vertically, horizontally, or, most commonly, both) may be sufficient to cause severe structural damage. A large quake, with its typical displacements in the fault zone of several feet, could demolish even a well-engineered building.

In addition, there is the problem of severe earthquake vibrations in a fault zone. Most structural damage to property is directly related to the intensity of shock waves in the ground, and ground shaking is usually very intense along the fault. Thus, even if there is no faulting, any building on or within the zone of a fault is exposed to very strong ground motions.

All of these facts have been known to geologists and soils and structural engineers for years. They have also identified and mapped the locations of many, but not all, of the most dangerous fault zones in the United States. In California, these are often called Alquist-Priolo (AP) zones, for the two state senators who legislated a bill in 1972 mandating the identification and mapping of active faults. Under this legislation, zone boundaries are set 500 feet away from most mapped faults and 200 feet for less significant faults. Sellers must inform potential buyers if the property for sale lies in an Alquist-Priolo zone. Although the Alquist-Priolo Act has had an impact on new construction, buildings continue to stand (and be bought and sold) right on top of fault traces.

People are commonly under the misapprehension that contemporary building codes in earthquake-prone areas take the danger of faults and faulting into consideration. It is true that modern building codes are written with the awareness of the high risk of earthquakes and set certain minimal standards of design and construction. However, in accordance with these building codes, almost all structures must meet the same standards (an important exception is that of hospitals, schools, and other critical facilities, which are designed to withstand higher forces). Permits for the construction of residential, commercial, and public buildings in areas subject to earthquakes are still issued only on the basis of these minimal earthquake-resistant standards. Except in California, the special geologic conditions and hazards of fault zones are not adequately recognized. Active faults in cities such as Seattle and Salt Lake City are still not recognized legally as dangerous locations for building.

The Hayward fault and the whole East Bay region of the San Francisco Bay Area provide a particularly instructive illustration of our past disregard of fault zones. The Hayward fault has experienced major faulting and causes small quakes that are felt throughout the Bay Area every few months. Yet even a partial list of the schools, important buildings, and places of public assembly very near the fault would fill a full page of this book. The newer of these structures, particularly the schools, were carefully designed and built for earthquake resistance. In many cases, however, particularly involving those buildings straddling the fault, this care will prove largely futile in a large quake. As for the hundreds of homes, businesses, and other buildings that were built with less knowledge and care, an optimistic earthquake scholar would have to predict a broad swathe of damage from San Pablo to Fremont in the event of a long surface rupture along the fault. See figure 2.

HOW FAULTING DAMAGES BUILDINGS

No building straddling a fault can withstand the abrupt surface ruptures and displacements of earthquake faulting without severe damage.

Vertical Faulting

Vertical faulting thrusts the ground surface upward, destroying the foundation above it, which, in turn, breaks the structural integrity of the building. This photo shows faulting damage as a result of the 1971 San Fernando earthquake. The structural damage of this house (above) was extensive and very costly to repair even though the thrust of the fault was small.

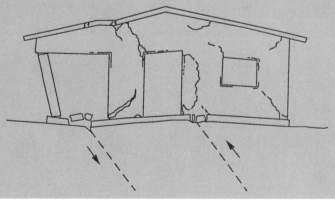

PEACE *of* MIND *in* EARTHQUAKE COUNTRY

Lateral Faulting

Lateral faulting splits the foundation and in larger displacements tears the building apart. An illustration of what extremely large lateral faulting could do to a foundation slab can be seen in the 1906 photograph (below) of a road that crossed the San Andreas fault near Point Reyes Station. The faulting displacement was more than 15 feet in this area near the epicenter of the earthquake.

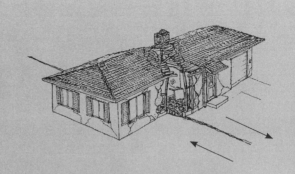

Tension Cracks

A network of tension cracks in the ground surface and pockets of slight landslide movements are both common earthquake effects in a fault zone passing through deep alluvial soil. The fissures or settlements will crack a foundation, and this damage inevitably results in fractures in the structural supports of a building. In this photograph of a house in San Fernando (right), downhill landslide displacements of a few inches opened up the narrow tension crack that passed beneath the house and continued down the slope behind it. The damage to the foundation and the frame of the house is not apparent here, but it was so severe that the house was declared unsafe for occupancy. The lower right photograph shows typical interior damage from these types of faulting.

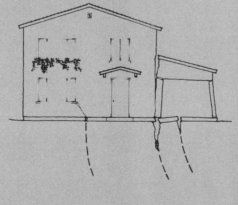

Where the Faults Are: Determining Fault Risk to Your Property

A look at a geologic map of California shows that none of the state's major population centers is very far from an active fault. In the greater San Francisco Bay Area, downtown San Francisco is only 9 miles away from the San Andreas fault. Oakland is about 15 miles from the San Andreas and is traversed by the Hayward fault as well. San Jose is bounded on the west by the San Andreas and on the east by both the Hayward and Calaveras faults. The Calaveras fault also cuts through some heavily populated suburban cities of the East Bay, while the San Andreas crosses residential Daly City and then skirts the suburban cities of the San Francisco Peninsula.

Central San Diego sits on top of the Rose Canyon fault. Northeastern Los Angeles is near the San Andreas fault, San Bernardino is astride it, and all of metropolitan Los Angeles is crisscrossed by numerous faults related to the San Andreas. Santa Barbara is near the Santa Ynez fault zone and its subsidiary fault traces. Sections of the Imperial Valley and lower Southern California are bisected by the Imperial fault. In addition, there are numerous other active faults in the state—near Eureka, in the Sacramento Valley, and near Bakersfield.

It is not just in California that densely populated areas sit close to faults. Numerous other large and assuredly active faults exist throughout the United States. Seattle is on the Seattle fault zone; Memphis, Tennessee, sits within a few miles of the locations of three of the most powerful earthquakes in U.S. history. Salt Lake City and many of Utah's largest towns are on top of or quite near the active and very large Wasatch fault, and a particularly treacherous fault zone—the East Bench—crosses Salt Lake City.

In addition to the question of location, there is also the problem of determining whether a given fault is active or inactive. The White Wolf fault, a short and relatively insignificant fault in the Arvin-Tehachapi area between Bakersfield and Los Angeles, was considered inactive for many years until on July 21, 1952, that theory was dramatically dispelled in a destructive magnitude 7.7 earthquake. Similarly, the 1971 San Fernando earthquake occurred on the same type of fault—one that had been dutifully plotted on the more detailed geologic maps and then largely ignored as inactive. More recently, the Coalinga earthquake of 1983, the Whittier quake of 1987, and the destructive Northridge earthquake of 1994 were all on unmapped faults. As you can see, the majority of Southern California's destructive earthquakes, to

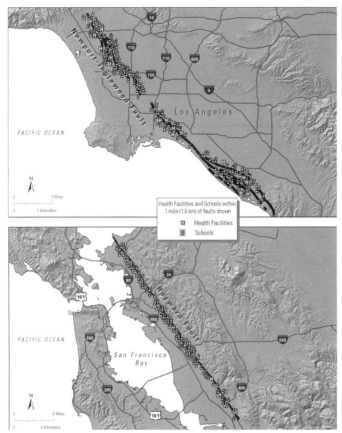

(Figures 1 and 2)

(Figures 1 and 2) In overall risk and potential for damage, the Newport-Inglewood fault through the Los Angeles area and Orange County is one of the most dangerous faults in Southern California. It bisects heavily populated areas and numerous industrial areas, such as Long Beach. It is also near numerous schools and hospitals as shown in figure 1.

Similarly the East Bay region of the San Francisco Bay Area includes one of the most densely populated fault zones in the United States. The Hayward fault, which bisects numerous large cities along the bay, has not caused a destructive earthquake since the area became heavily settled, but several destructive quakes occurred on the fault in the nineteenth century. The fault is highly active, with frequent small tremors and creep zones, so most earthquake experts rank the fault among the top five most dangerous earthquake zones in the United States. Note that the schools and hospitals marked on the map represent only a few of the many important structures near this fault zone. A large earthquake could cause damage easily exceeding $150 billion.

(Figures 3–5) The San Andreas fault zone on the west side of San Francisco Bay is one of the most famous—and dangerous—fault zones in the world. The danger is graphically illustrated in these photographs, taken in the western part of Daly City in 1956 and 1966 and again in 2007. The sag pond, landslide, and other geologic features of the fault zone in the top picture have been obliterated by the landfill and concrete of a suburban tract. Heavy property damage and some loss of life are a certainty when the next big earthquake strikes along this portion of the San Andreas. Note the empty lots in the lower photograph just above the landslide area.

take just one region of the United States, during the last several decades were on unmapped faults. Thus, even if your property is not near a known active fault, it may be near an unknown fault that could cause a strong local earthquake. Therefore, no matter where you are in earthquake country, you should take earthquake risk management seriously.

It is impossible at the present time to predict the future behavior of known faults. In addition, new faults, or at least those new to science, will continue to appear abruptly as they shake and rupture the ground in an unexpected tremor. Only one thing is certain—earthquakes and associated faulting will occur in the United States, and there is nowhere, particularly in California, the Pacific Northwest, western Alaska, and the area of New Madrid, Missouri, that one can go to completely escape the risk of tremors. Other states are generally less active, but they can be more dangerous because we know less about their risks and build weaker buildings there.

To establish the earthquake risk of a given site, you will need to locate your property on a geologic map that shows local active and inactive faults. Such maps are essential because faults are not always obvious to the eye. The break always occurs initially deep below the surface of the earth, and it may never surface with ruptures or displacements, even during violent earthquakes. In addition, surface evidence of faulting can be absorbed and hidden by excavation and landfill, under the natural deposits of deep alluvial soils, or below the ocean or a lake surface. These maps are now available online, as discussed in appendix A.

Certainly, neither individual property owners nor developers are usually anxious to divulge the information that a fault runs under or near their property. In California, as mentioned earlier, the Alquist-Priolo Act requires the seller of a property in a known active fault zone to divulge that information to the prospective buyer. Until regulatory laws and building codes are reformed elsewhere in the United States, however, the prospective buyer or builder and the present owner can rely only on professional consultants and/or information available in the maps of the U.S. Geological Survey and other state and local agencies.

Another step necessary to establish the earthquake risk of a given site is to determine the history of the nearest fault—the frequency, magnitude, intensity patterns, and displacements or surface ruptures of past quakes. All of these considerations are essential in weighing the risks of living, buying, or building near (not in) a fault zone. You can begin your initial research in the appendix.

(Figure 3) 1956

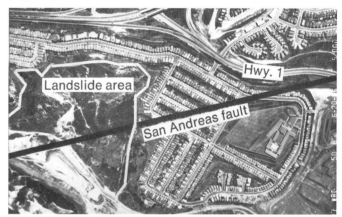

(Figure 4) 1966

(Figure 5) 2007

WHEN WILL THE NEXT BIG EARTHQUAKE HIT THE SAN FRANCISCO BAY AREA?

There is a 62 percent probability that at least one earthquake of magnitude 6.7 or greater will occur on a known or unknown San Francisco Bay region fault before 2032. After a century of study by geologists, many faults have been mapped in the region, but not all faults are apparent at the surface—quakes continue to occur on unknown faults. The probabilities for some of the major (and known) active faults are shown on the right.

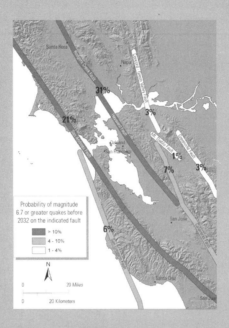

Estimated consequences of a major earthquake (in this case a repeat of the 1906 M7.9 quake) striking the Bay Area are represented below.

- Instantaneous deaths: 3,400 if the event occurs during the day; 1,800 if at night
- Seriously injured: 12,500 if the event occurs during the day; 8,000 if at night
- Number of households displaced: 250,000 households or 600,000 people
- Direct economic losses (including business interruption): $120 billion
- Percentage of San Francisco commercial buildings with extensive or complete structural damage: 40 percent

Another way to estimate earthquake loss in the Bay Area is to look at losses from the 1995 Kobe, Japan, earthquake. Kobe and the Bay Area have similar geology and population densities. In Kobe, the following losses were recorded:

- 6,400 people were killed.
- $150 billion in direct losses.
- 400,000 buildings damaged.
- 85 percent of schools and hospitals suffered major damage.
- 850,000 homes lost gas service for two to three months (during the winter).
- Restoration of water and wastewater services to nearly 1.3 million households took up to four months.
- Three major freeway systems collapsed.
- The Port of Kobe was heavily damaged, with repairs taking up to a year to complete.
- Chemical and steel manufacturers were out of operation for several months.
- 80 percent of Kobe's 2,000 small- to medium-sized businesses failed.

WHEN WILL THE NEXT BIG EARTHQUAKE HIT CALIFORNIA?

As stated in a recent comprehensive study, there is a 99.7 percent chance that a major earthquake (magnitude 6.7 or greater) will hit California in the next thirty years. For the optimists out there, there is still a 0.3 percent chance that the state will get off the hook.

The likelihood of at least one magnitude 7.5 or greater in the next thirty years is 46 percent. The tables below show regional earthquake probabilities (left) and statewide earthquake probabilities (right). The figures below show some of the major faults running through the two most vulnerable areas in California—the San Francisco Bay Area and Los Angeles metropolitan area.

REGIONAL 30 YEAR EARTHQUAKE PROBABILITES

MAGNITUDE	SAN FRANCISCO REGION*	LOS ANGELES REGION*
6.7	63%	67%
Magnitude	Northern California**	Southern California**
6.7	93%	97%
7	68%	82%
7.5	15%	37%
8	2%	3%

* Probabilities from UCERF for the San Francisco region are nearly identical to the previous results from WGCEP 2003.
** These probabilites do not include the Cascadia Subduction Zone

STATEWIDE EARTHQUAKE POSSIBILITIES
The numbers represent current best estimates. As earthquake science progresses, these probabilities will change. Actual repeat times vary considerably and only rarely will be exactly as listed in the table.

MAGNITUDE	30-YEAR PROBABILITY OF ONE OF MORE EVENTS GREATER THAN OR EQUAL TO THE MAGNITUDE	AVERAGE REPEAT TIME (YEARS)
6.7	>99%	5
7	94%	11
7.5	46%	48
8	4%	650

*Not including Cascadia Subduction Zone

San Francisco Bay Area

Los Angeles

THE HAYWARD FAULT: THE ODDS AREN'T GOOD

The Hayward fault, a branch of the San Andreas fault system that runs through San Francisco's East Bay, is an offspring even more dangerous than its treacherous parent.

Most engineers and seismologists believe that the Hayward fault could produce the most destructive quake in the Bay Area. That's because it runs through the most densely populated and oldest cities of the East Bay and is astride Silicon Valley. It also has the highest probability, based on history, of causing the next magnitude 7+ quake. The U.S. Geological Survey believes that there is at least a 31 percent chance of a major earthquake on this fault before 2032.

The Hayward fault passes through virtually every city on the eastern shores of San Francisco Bay before it enters the bay at Point Pinole near San Pablo. It emerges from the bay in Sonoma County and continues toward Petaluma along the Petaluma Valley. The fault then appears to merge into two other fractures, the Rodgers Creek and Healdsburg faults, which continue north past Santa Rosa to Healdsburg.

The Hayward fault has caused several destructive quakes, including the Hayward earthquake of 1836 (estimated magnitude 6.8), which was one of the largest ever to occur in the Bay Area. According to a recent study of that quake, fissures opened along the fault from San Pablo to Mission San Jose (now part of Fremont), and the shaking caused havoc in the settlements of Santa Clara and Monterey.

In another great quake, in 1868, also with an estimated magnitude of 6.8, the fault ruptured for about 20 miles, from Warm Springs (now part of Fremont) to the vicinity of Mills College in Oakland. Horizontal displacements were up to 3 feet, and every building in the village of Hayward was either severely damaged or completely demolished. Numerous structures in San Francisco, particularly in the filled areas of the bay, were also destroyed or damaged. The intensity map for this quake is shown below.

Another earthquake, on October 7, 1915, was centered in the vicinity of Piedmont, where most of the damage occurred; the shock was felt as far as Sebastopol and Santa Clara. And on May 16, 1933, the fault erupted again, in the vicinity of Niles and Irvington (now incorporated into Fremont), where all chimneys were thrown down and numerous dwellings damaged. The most recent damaging earthquake along the Hayward fault occurred on March 8, 1937, in the Berkeley-Albany-El Cerrito area.

Estimated financial losses for a repeat of the 1868 earthquake are likely to exceed $150 billion. This does not include losses due to business interruption.

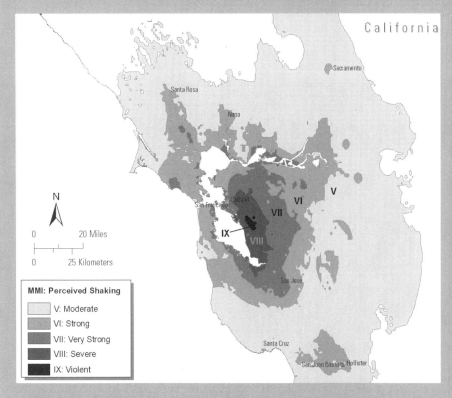

1868 Hayward Fault Earthquake

How Far from a Fault Zone Is Far Enough?

Unless your property is within an active fault zone, the earthquake hazard to your property will vary, and simple distance is not always the most important factor. Generally speaking, of course, the farther you are from a fault, the better off you will be. However, the geologic foundation of your site also plays an important role. As you will learn in the next chapter, certain soil foundations intensify the shock waves of a quake. Certain soils are also prone to severe settlement during a quake—an action that can cause serious damage. In a large earthquake in which the shaking intensity remains strong many miles from the fault, a house or building on one of these unstable geologic foundations may suffer more damage than one on stable ground very near the fault zone.

In making a decision about the relative risks of various locations, property owners or buyers should consider the sites' geologic foundations and the past history and future prospects of the nearby fault. Generally, anyone living in coastal California, Oregon, Washington, British Columbia, Alaska, Utah, or the New Madrid area should assume that the location is too near a potentially explosive fault zone and thus should give primary consideration to the geologic foundation of the property, the strength of the building, and the potential need for insurance coverage.

There is one further consideration. The surface ruptures of faulting are almost always restricted to the relatively narrow area immediately adjacent to the fault line. Some faults, however, do not follow this linear pattern. Instead, they tend to fracture the ground surface over a broad area, sometimes extending for hundreds or thousands of feet.

As long as a building is not located within a fault zone, it can survive an earthquake if (1) it is on a stable geologic foundation, and (2) it is constructed or reinforced to be resistant to earthquake forces.

What to Do About a Property in a Fault Zone

If you own property in or near an active fault zone, you should consult a geologist to determine, first, whether the fault has demonstrated surface ruptures, displacements, and/or creep in your area in the recent past. If not, there is a chance that the fault beneath or beside you will remain true to its history: an earthquake on that fault could subject your home only to intense vibrations and not to the foundation-splitting displacements of faulting. In this case, a further consideration is important to the survival of your property: the strength of your building's design, structure, and materials.

(Figure 6) Faulting from the 1987 New Zealand earthquake extended within 50 feet of a single-story wood-frame house. Yet there was no ground disturbance underneath the well-engineered home, and it was not damaged.

(Figure 6)

UNDERRATED FAULTS THROUGH MAJOR METROPOLITAN AREAS

Justifiably, the most engineering attention, as well as the most media attention, has gone into determining and discussing the hazards from well-known faults in the Los Angeles and San Francisco areas. These are America's largest metropolitan areas bisected by numerous and well-known active faults.

Much less attention has gone into publicizing and reducing the less-frequent risk from large active faults in the other earthquake prone metropolitan areas of earthquake country. The accompanying photographs illustrate the locations of major active faults with respect to the denser centers of some of these areas, which include San Diego, Seattle, Salt Lake City, and Portland. Although they are not shown, we also include Saint Louis, and Memphis in this group—their major threat is not located next to downtown, but it is close enough.

The central business districts of all of these cities are populated by older, larger buildings that were built without any earthquake design criteria. Some of the very tall modern buildings were, until the late 1970s, designed for what we now consider insignificant earthquake criteria. Unlike in most of California, most of these buildings have not been strengthened during the past two decades. Some of these high-occupancy buildings are collapse hazards. It is foolhardy to live or work in such structures—they should be carefully evaluated and strengthened to meet at least life-safety criteria.

All of these areas also share a common problem caused, to a large degree, by earthquake engineers and building officials who are trying hard to keep down the cost of construction. Modern building codes are effectively based on three considerations: (1) the size of the possible earthquake, (2) the probability of that earthquake occurring during the life of the building, and (3) the needed strength of the building to withstand the shaking. This probabilistic approach has a major problem. The issue is simple—if the earthquake does not strike all that often, then we design for the more probable, lower-strength earthquake in order to make the construction more affordable. Why really worry that much about an earthquake, such as one on the Seattle fault, that happens, at best, only every two or more hundred years. The probability is obviously low. And that is what the codes require—to design to standards that are much lower than those required for Los Angeles. The problem caused by this approach is also simple—none of the buildings (the concern here is with the larger or very large buildings) are designed to withstand the large and very infrequent earthquake. They simply do not have the strength to withstand the worst-case earthquake. But that earthquake can happen tomorrow. In effect, unless you have taken the appropriate risk-management steps, you may be safer living and, especially, working in Los Angeles and San Francisco.

San Diego, California

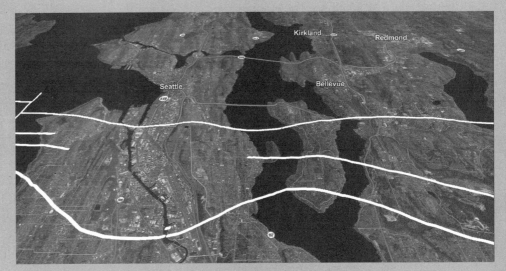
Seattle, Washington

Salt Lake City, Utah

Portland, Oregon

1989 Loma Prieta earthquake, Marina District, San Francisco

1993 Hokkaido, Japan, earthquake, Island of Okushiri

CHAPTER 3

Rock, Soils, and Foundations

During the 1906 San Francisco quake, which occurred in the early morning, some people living on top of the city's famous hills slept through the enormous tremor, and numerous buildings on these bedrock hills survived the quake without serious damage (although they were later consumed by the destructive fires). Meanwhile, in homes and buildings atop landfill along the bay and alluvial soils between the hills, people were thrown out of bed by the shock and found themselves unable to get up on their feet during the forty to sixty seconds that the motion lasted. Most of the buildings that collapsed were also in these areas.

The same thing happened at 5:04 P.M. on October 17, 1989, during the Loma Prieta earthquake, centered more than sixty miles south of downtown San Francisco. Old houses and other buildings on the rock of Pacific Heights suffered no substantial damage, whereas many wood-frame buildings on old landfill in the Marina, just a few blocks away, collapsed or were severely damaged. The other spectacular damage, in the city's South of Market area and along Interstate 880 in Oakland, where 1.5 miles of a two-level freeway structure collapsed, also occurred on very soft soils and old, poorly compacted fills.

Quake effects are sharper in alluvial and landfill locations because the intensity of vibrations increases as earthquake waves enter a thick layer of soft soil or the less dense mixtures of soil found in most landfills. These soft, unstable soils act much like jelly in a bowl, responding to and then amplifying earthquake motions, so that rapid, small-amplitude vibrations in the bedrock are transformed into slower and more damaging large-amplitude waves. The chaotically undulating motions of these waves can be devastating at the surface, particularly in landfill and water-saturated soils.

An earthquake-hazard map of California, Alaska, and the other earthquake-prone regions of the United States would show that, aside from the fault zones, the highest-earthquake-hazard areas are those with natural alluvial soils and man-made landfills in valleys and near the coast and bays. Hills and mountains, which are composed mainly of bedrock with a thin soil surface, would appear on the map as the lowest-hazard areas.

A site on the bedrock of a hillside is safer than most others, unless the building is in a landslide area or the builder has cut into the slope and then rests on the loose cuttings for landfill that lacks the proper grading, compacting, and drainage. Post-earthquake studies have consistently shown that structures built on rock near the fault or epicenter of an earthquake fare better than much more distant buildings on soft soils.

A highly detailed soil map would show wide variances in risk from one site to another in a city or a neighborhood, since urban areas are inevitably composed of a wide variety of natural and man-made soil foundations, including old and forgotten landfilled watercourses, sand dunes, and water-saturated mud disguised by housing and office developments. During the 1971 San Fernando earthquake and again in the 1994 Northridge earthquake, buildings within the small area of the Caltech campus in Pasadena, many miles from the fault, recorded ground accelerations that varied by a factor of two. This large difference in ground-shaking intensities is explained by the fact that various

(Figure 1)

(Figure 2)

(Figure 3)

(Figure 4)

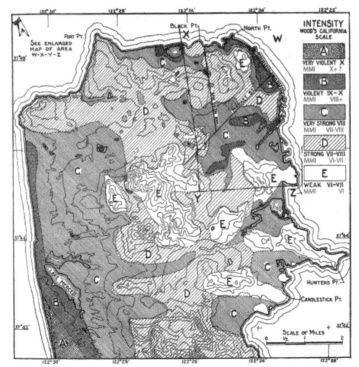
(Figure 5)

geologic foundations beneath the campus responded to the earthquake in considerably different ways. The same effects were observed in San Francisco in 1989. Ground motions recorded in the rock of Pacific Heights, a hill a few hundred yards from the Marina, were one-quarter to one-fifth as strong as motion in the fill of its less fortunately located neighbor.

The importance of geologic foundations in minimizing earthquake damage can be demonstrated with a profile of the destruction caused by the 1906 quake. The intensity map shown here illustrates the distribution of ground and building damage in San Francisco. The stability of rock foundations is clearly shown by the lighter structural damage on the hills—Telegraph, Russian, and Nob hills, Pacific Heights, Twin Peaks, Hunters Point, and Potrero Hill. On the other hand, the flat alluvial areas between the hills were hit hard by the quake, with moderate to severe vibrational intensities and structural damage. Suffering the greatest amount of damage were the business district in the vicinity of the Ferry Building and most of the lower Mission district, which were built on landfill atop already unstable bay mud. (The Marina area did not exist then.) Although these areas were farthest (some 9 or 10 miles) from the fault in San Francisco, the land heaved and then settled drastically under the force of the shaking, creating havoc among the light wood-frame structures. In addition, numerous fissures opened in this unstable ground, allowing the high water table to surge above the surface.

The sandy beach and dune areas along Lake Merced and Ocean Beach were also severely shaken, partly because the fault's proximity and partly because of unstable water-saturated soils. At the time, this region was sparsely settled and damage was light. Today, an earthquake of the same magnitude would strike a densely settled area with many older homes and buildings.

In that same earthquake, Santa Rosa suffered proportionally even greater damage than San Francisco did because of its geologic foundations, even though the city is more than 19 miles from the San Andreas fault—twice as far as the farthest point of San Francisco. Its central business district, which sits atop an alluvial plain, was almost completely destroyed. The same comprehensive damage affected San Jose, which is on a similar soil foundation nearer the ruptured fault. It is estimated that these and other localities relatively distant from the fault experienced vibrational intensities equal to or sometimes greater than those in the most affected parts of San Francisco. We expect that history will repeat itself in the next large earthquake.

The same lessons were repeated in San Francisco during the 1989 Loma Prieta earthquake. This time, however, sensitive instruments called accelerometers recorded ground motion, so we can make direct, numerical comparisons. These instruments record acceleration as a fraction of gravity, which is notated as 1.00g. The strongest acceleration, 0.64g, was recorded at Corralitos near Santa Cruz, about 4 miles from the epicenter. The rock hills of northern San Francisco, at Pacific Heights and Telegraph Hill, recorded 0.06g and 0.08g, respectively, at about 61 miles from the epicenter. There was no significant damage there. The nearby Presidio, on softer ground, recorded 0.21g. The probable acceleration in the landfilled Marina, which is sandwiched among the above three areas, was about 0.25g to 0.35g, or about four to six times stronger than that in nearby Pacific Heights. The recorded acceleration in the soft, filled land of San Francisco International Airport recorded 0.33g at about 50 miles from the epicenter. Only 0.14g was recorded in nearby San Bruno on firm ground. The same comparisons can be made in Oakland, where again accelerations on soft or filled ground were several times stronger than those on rock.

A similar concentration of damage was seen in soft-soil areas of Los Angeles after the 1994 Northridge earthquake. Damage farther from the epicenter was generally attributed to either soft soils and/or poor building construction.

In addition to amplifying ground acceleration, soft soil tends to change the frequency characteristics of ground motion, which can have its own damaging effect on certain types of structures. Taller, more flexible structures in particular tend to be more affected by earthquakes when supported by soft soils.

(Figures 1 and 2) The building failures here are typical of the vibrational and settlement damage in alluvial and "reclaimed" bay lands of San Francisco in 1906. Note that some of the buildings, which were either better constructed or located on sounder soil foundations, survived the quake with little apparent damage.

(Figures 3 and 4) Similar scenes occurred in San Francisco in 1989 on the types of soil where engineers had predicted damage. These scenes are from the Marina district. Both of these structures were four-story apartment buildings with garages on the ground floor. Similar buildings on firm soils nearby suffered no damage.

(Figure 5) This intensity map of the city of San Francisco during the 1906 earthquake indicates the widely varying effects of the shock in different soil foundations. The most violent areas were the ocean beaches of the city, which are soft soil and are also nearest the San Andreas fault. Other equally violent areas farther from the fault are associated with the poorest ground conditions: land-filled swamps, water channels, bayside lands, and deep natural soil foundations (alluvium) between the hills. Buildings on the bedrock of the numerous hills of the city were least affected by the powerful earthquake. This pattern was repeated for the much smaller 1989 earthquake.

NATURE'S GREAT FEEDBACK LOOP

Imagine grabbing the top of a tall, slender building, pulling it to one side, and releasing it, allowing the structure to vibrate from side to side (see image on the right). The time that the structure takes to complete one vibrational cycle is called its natural period of vibration. For a tall building, the period will be relatively long, maybe several seconds. Now imagine doing the same thing to a short and stiff building. The natural period of vibration will be shorter—usually much less than a second. The difference is due to the relative flexibilities of the two structures. The tall building is relatively flexible, while the short, rigid building is stiffer. Variations in mass (weight) can have similar effects on the natural period of vibration.

Now consider the vibration of the ground due to an earthquake and how this affects a structure. If the ground vibrates at a frequency that is similar to that of the structure, then resonance will occur. Anyone who has pushed a child on a swing understands the concept of resonance. If we push at a frequency that is the same as that of the child swinging, then the swing becomes higher and higher. It is the same with an earthquake and a structure. If the ground (the swing pusher) moves at a frequency that is similar to that of the structure (the swing), then the response (the swing height) becomes greater.

During an earthquake, firm ground vibrates at a frequency that is close to that of short to midrise buildings, thereby causing greater damage to such structures. Soft ground, on the other hand, vibrates at a lower frequency, and thereby has a greater effect on tall or long structures.

Most tall, flexible buildings are connected through piles to rocks and are therefore relatively safe as compared to short, stiff structures on firm soil or tall, flexible structures on soft, flexible soil. Such safer structures include most tall buildings in downtown San Francisco and Los Angeles and long-span bridges such as the Golden Gate Bridge.

The effect of soft soil on building response was graphically illustrated during the 1985 Mexico City earthquake as per the figure below. The collapse of many tall commercial and apartment buildings that were not founded on rock was caused by soft soils that changed the frequency of the ground motion to match that of the doomed structures.

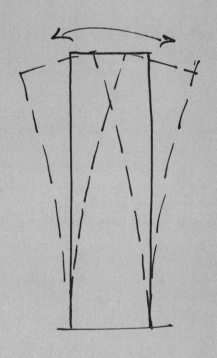

Mexico City, 1985

(Figure 6) This dramatic example of the effects of liquefaction on sandy or water-saturated soils occurred in Niigata, Japan, during the large earthquake centered near there in 1964. The apartment buildings suffered moderate structural damage, but soil liquefaction left some of them tilted at 80-degree angles. Some occupants evacuated by walking down the buildings' faces. The same type of damage occurred in the M6.8 2007 earthquake that struck the Niigata region of Japan.

(Figure 7) The same effects were observed during the M7.7 1990 Central Luzon, Philippines, earthquake.

Soil Liquefaction

Soil liquefaction is a very common effect of earthquakes in low-lying coastal areas and wherever sandy soils and high water tables exist (near bays, lakes, rivers, deltas, and marshlands, for example). The compaction of sandy soil by earthquake vibrations causes water pressure in the soil to increase, pushing apart soil particles to the point that they can't hold weight. The result is a kind of quicksand. This is similar to the experience of standing in a tidal area at the beach, moving your feet back and forth, and finding yourself sinking into the sand.

By far the most dramatic example of liquefaction occurred in the coastal city of Niigata, Japan, during a magnitude 7.4 earthquake in June 1964. Although the epicenter of this major earthquake was about 35 miles from the sea-level city, liquefaction developed over large sections of the city, and numerous buildings, automobiles, and other heavy objects gradually settled into the newly formed quicksand. Many apartment buildings settled several feet and tilted at such a rakish angle that occupants made their escape by walking down the walls. The same type of damage happened in August 2007 when a smaller, magnitude 6.8 earthquake struck the town of Kashiwazaki, Japan, just south of Niigata City.

The Marina district of San Francisco suffered extensive liquefaction, as expected, in the 1989 Loma Prieta earthquake. The area had been carefully filled for the 1915 Panama Pacific Exposition and built up in the 1920s and 1930s before modern soil compaction techniques were developed. Liquefaction extensively damaged hundreds of houses and apartment buildings in the area. Note, however, that the deadly collapses of four-story apartment buildings occurred primarily because of structural weaknesses, as is discussed in chapter 7.

The Port of Kobe, Japan, suffered major damage due to liquefaction during the 1995 earthquake. Large sections of the wharf and warehousing areas sank, and approximately fifty cranes were destroyed. The Port of Kobe is the sixth largest in the world, and damage resulted in widespread business interruption to global businesses. With shipments being diverted to Yokohama, Osaka, and South Korea after the earthquake, the Port of Kobe has still not recovered to its pre-earthquake shipment volume.

Studies have shown that numerous areas of California, Alaska, Utah, and Washington are susceptible to equally spectacular effects—in particular, landfill areas in former delta or marsh regions, such as sections of Newport Beach, Long Beach, San Diego, Livermore, the area around Puget Sound, and Vancouver. Other areas include the filled or diked lands bordering San Francisco and other bays, ocean beach developments, and landfill sites near the mouths of rivers in Puget Sound and throughout the Los Angeles area. If you own, purchase, or build a structure in such areas, seek the advice of a soils or geotechnical engineer regarding the possibility of liquefaction. The problem, if it exists under your building, is expensive to correct.

(Figure 6)

(Figure 7)

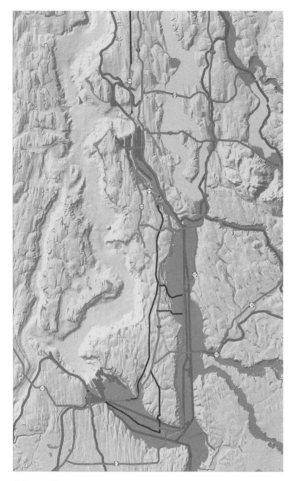

(Figure 8)

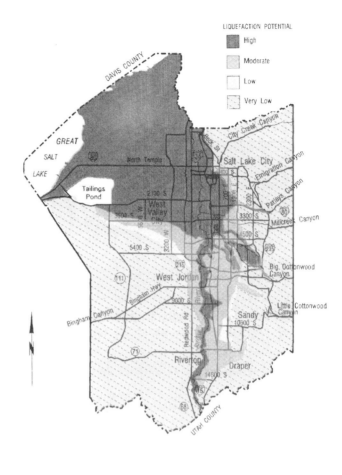

(Figure 9)

(Figure 10)

(Figures 8–10) These illustrations show areas susceptible to liquefaction for Seattle (figure 8), Salt Lake City (figure 9), San Francisco (figure 10).

(Figures 11–13) Port damage due to liquefaction and settlement from the M6.9 Kobe earthquake of 1995. Japanese engineers were convinced before the earthquake that they had controlled this problem with massive prefabricated concrete retaining wall structures and with other innovative earthquake designs previously untested in earthquakes. The port structures were in fact not designed well and became a massive engineering failure that destroyed most of one of the world's largest ports.

(Figure 11)

(Figure 12)

(Figure 13)

Problems with Landfills

Like any soft soils, landfills—particularly hydraulic sand fills and other poorly compacted fills typical of construction before the mid-1950s—pose a much higher risk for earthquake damage than most other soil foundations. Indeed, improperly engineered landfill is often significantly more damaging to structures than even some of the poorest alluvial soils; the fill is often loose and insufficiently cohesive and will shake and settle drastically when the shock waves of an earthquake pass through it. Landfills are also frequently full of organic matter, which decays and creates voids and weak spots prone to settlement. Old, filled refuse dumps, for example, are among the most damage-prone sites in earthquake country because of the excessive amount of organic matter beneath their surface. Such sites, particularly those close to active faults, should be avoided when you are property hunting. If you already own such property, relocation or extensive reinforcement and full insurance coverage should be the rule.

Modern methods of landfill engineering, which include careful compaction and selection of fill materials, have reduced earthquake hazards associated with landfill foundations. During the 1957 and 1989 San Francisco earthquakes, for example, the well-compacted fill under the newest subdivisions of Daly City was apparently not a significant factor in the moderate structural damage to homes in the area, even though some of it was 35 feet deep. However, the 1957 earthquake was a minor one, and the 1989 quake was distant and not strong enough to subject these landfill sites to great vibrational forces. Whether the filled areas are sufficiently well engineered to protect structures during a large quake remains to be seen. The extensive damage in Kobe is certainly thought provoking. Good engineering does not guarantee good results, primarily because good design is not necessarily followed by good construction. If the contractor decides to save money or take shortcuts, and the engineer does not spot the sloppy work, the quality of the fill will be reduced and liquefaction or settlement could occur. As a rule, a structure in a fill area runs a much higher risk of damage in an earthquake because the ground beneath it may fail even if the structure itself is built properly. In Kobe, what many Japanese engineers thought to be well-engineered, modern fills performed very poorly, leading to severe and extensive damage to thousands of buildings and industrial structures.

Problems of Bayside, Riverside, and Watercourse Sites

The flat alluvial lands along the shores of San Francisco Bay, the Santa Ana and Sacramento River deltas, Puget Sound, and numerous other bayside and riverside sites present a special earthquake problem. These areas are composed largely of thick deposits of a soft, silty clay that is highly compressible and unstable and has a high water content—all poor characteristics in high-intensity earthquake regions. The ground motions of an earthquake are amplified by this soft, water-saturated soil and can cause compaction of the clay and settlement of the ground surface. In addition, the high water content of such soils can induce liquefaction.

As we have noted, ground settlement can be as damaging to a building as fault displacement. For example, all comprehensive engineering reports on areas along San Francisco Bay agree that, in large earthquakes, intensities and damage could be as high in bayside developments as they would be in the fault zone itself. Housing developments such as Redwood Shores, Bay Farm Island, Foster City, and marina homes and apartments in Emeryville, Oakland, and Alameda face much greater risks because they are largely on new landfill over bay mud. Some of these developments also face the risk of flooding after an earthquake, because the perimeter dikes that protect them from high tidal waters are highly susceptible to earthquake damage and collapse.

Port facilities are at high risk for this same reason. The ports of San Diego, Long Beach, Los Angeles, Oakland, Puget Sound, and Vancouver are highly susceptible to major damage, which would have a ripple effect on businesses throughout the United States and the world, as happened following damage to the Kobe port in 1995.

Before you buy a home, rent an apartment, or invest in any facility next to water, ask the developer for a full report on the precautions that were taken to protect the buildings from severe earthquake damage. If necessary, have an experienced engineer perform an evaluation. If you are building in these areas, see that the principles of structural earthquake resistance outlined in later chapters of this book are fully incorporated into the plans for your structure. The moderate additional construction costs are very likely to prevent serious damage in an earthquake, and the supplemental reinforcement will always result in a better, stronger, more durable structure.

Riverside and old watercourse sites face essentially the same problems as bay lands. Buildings located near or along present and former rivers, creeks, marshes, and so forth usually have a much higher than average earthquake risk. The strongest shaking in the destructive 1933 Long Beach earthquake occurred near the coast adjacent to the mouth of the Santa Ana River. An earthquake near Puget Sound in 1965 caused considerable damage to buildings in low-lying and filled regions along the Duwamish River in the Seattle area. The residential and commercial developments of Harbor Island, at the mouth of the Duwamish, were also hard hit. This was again the case in the 2001 Nisqually, Seattle, earthquake: there was much liquefaction-induced damage in the industrial area along the Duwamish north of the King County Airport and south of downtown Seattle. Much, if not all, of Harbor Island was man-made, and its soil is seismically unstable by any standard.

The 1906 shock in San Francisco caused spectacular land failures along several filled creek beds within the city. During that same quake, much of the ground along the Salinas River in Monterey County lurched and settled severely, completely destroying small structures in the area of failure. Numerous buildings along the Russian River, north of San Francisco, suffered similar fates.

Numerous San Francisco buildings over and along old creek channels and swamp areas were destroyed or severely damaged by large ground settlements during the 1906 quake, and more were damaged in the Mission district in the 1989 temblor. Another,

(Figure 14)

(Figure 14) Houses along waterways, particularly when built on filled land, face higher risks in an earthquake.
(Figure 15) Houses on top of ridges appear to experience much stronger shaking due to very localized amplification of ground motion.
(Figure 16) The Madigan Ranch house, a few miles from the 1989 Loma Pieta (San Francisco) earthquake and situated on a ridge, experienced much stronger ground motion than its neighbors on flatter ground. Its cripple stud lower floor walls collapsed.

most instructive example of such localized failure occurred in eastern Turkey in a magnitude 7 earthquake in 1966. A regional school campus had been built across a former river channel, and the majority of buildings actually on the old channel completely collapsed because of amplified earthquake motions in the soft, unstable soil. Meanwhile, similarly constructed buildings on a higher gravel bench above the old river channel survived with only slight or no damage.

Higher Risks Along Cliffs, Ridges, and Hillsides

Ocean cliffs near large faults, such as the San Andreas in Northern California, the Santa Ynez near Santa Barbara, and the Newport-Inglewood fault zone in the Los Angeles area, present special risks during earthquakes. Because the cliffs are not supported by ground and rock on one side, they experience more earthquake motion than occurs in the ground some distance inland. In addition, as shock waves emerge from the ground, they reflect off the cliff face and are further amplified. The residential Westlake Palisades section of Daly City suffered the highest damage of any area during the minor San Francisco earthquake of March 1957. Earthquake experts relate the greater damage to the sea cliff that bounds this development on the west. Earthquake waves, rising almost vertically from the nearby San Andreas, were reflected and intensified by the steep cliff. The same effect was observed in Aptos, just south of Santa Cruz, in the 1989 Loma Prieta shock, where in general, houses along the cliffs suffered significantly more damage than nearby houses inland from the beach. Structures along such cliffs should be designed for the highest earthquake forces, particularly if the cliffs are near a major fault zone.

Houses on ridges are also exposed to higher risk. Much more damage to such homes has been observed in strong earthquakes, including those in San Fernando (1971), Morgan Hill in Northern California (1984), Chile (1985), Loma Prieta (1989), and Northridge (1994). The energy of earthquake waves appears to be trapped within the peak of the ridge, causing great amplification in a very local area—which leads to much larger forces on the buildings. Therefore, buildings on ridges need to be designed to the highest earthquake standards, comparable to those used for buildings within fault zones.

Certain hillside and hilltop developments should also be approached with great caution. Steep hillside sites are often graded and filled; if the fill is poorly compacted, these sites can be very risky. A moderate quake (or even a heavy rain) may cause such a fill to slip, taking the structure down the hill with it or severely damaging the building's foundation. There is also the danger of a poorly graded or supported cut above the house surging down upon the structure.

Any substantial cracks in the foundation, driveway, sidewalks, or patio of a building on a graded hillside or an apparently flat hilltop may indicate an inherent weakness in the site. If you notice such cracks around your home or building or suspect the existence of localized fills in your development or site, call upon a geotechnical engineer to aid you in investigating the records and the nature of your soil foundation. If you are shopping for a hillside or hilltop home or building, look for evidence of geologic instability. It is very costly and therefore seldom worthwhile to try to shore up a weak and unstable soil foundation.

(Figure 15)

(Figure 16)

(Figure 17)

(Figure 18)

(Figure 17) The heaviest building damage and the highest earthquake intensities were recorded along the ocean cliffs of Daly City during the San Francisco quake of 1957. The cliffs reflected and magnified shock waves emerging from the San Andreas Fault.

(Figure 18) Similarly, the 1989 Loma Prieta earthquake caused failures of sea cliffs from Pacifica, 50 miles north of the quake's epicenter, to Monterey, 30 miles south. The earthquake further undermined the previously exposed and endangered foundations of this apartment building in Capitola, near Santa Cruz. Six units overhanging the cliff had to be removed.

(Figure 19) Prior to the 2004 Indian Ocean tsunami, little or no attention was paid to a worst-case-scenario tsunami that could completely devastate an entire community, as happened to many locations in Indonesia and elsewhere.

(Figure 20) This image shows column failures in two-story reinforced concrete units on Khao Lak beach, as a result of the 2004 Indian Ocean tsunami.

Tsunamis

Tsunamis, seismic sea waves, are caused by faulting or other abrupt ground movements on the ocean floor or shore during large earthquakes. Sudden vertical displacement of the ocean floor can cause vast portions of the ocean to move vertically, resulting in a series of large sea waves. Tsunamis can also be generated by landslides or volcanic eruptions. Not all earthquakes trigger tsunamis, especially if the earthquake is associated with lateral ground displacement as opposed to vertical displacement. In the open ocean, tsunami waves are not much above normal height, but they move at very high velocities, sometimes reaching 400 miles per hour, and when they approach a shoreline, the offshore slope can raise them higher than 50 feet.

Tsunamis' devastating effects were graphically illustrated in December 2004 when a magnitude 9.1 earthquake in the Indian Ocean off the northwestern coast of the island of Sumatra, Indonesia, caused massive tsunamis waves that killed hundreds of thousands of people in several countries across thousands of miles. Tsunamis, and the corollary flooding and receding waters, are very destructive to structures.

There are two types of tsunamis that could affect the United States: local and distant. Local tsunamis involve earthquakes off the U.S. coast, while distant tsunamis originate as far away as Japan, Chile, and Russia. Local tsunamis present the greatest threat because they provide only a few minutes' warning (or less) before hitting shore. Distant tsunamis are a lesser threat because there is more time to notify residents. The greatest risk of tsunami is on the West Coast and in Alaska and Hawaii. On the East Coast and in the Gulf of Mexico, the risk of tsunami is less but not insignificant. Earthquakes in the Caribbean triggered by the Puerto Rico subduction zone caused moderate-sized tsunamis to hit Puerto Rico and the eastern seaboard in the past century.

Buildings along the west coast are not designed for the impact of tsunamis. If a building is within the inundation zone, the code requires nothing for protection from the impact of the waves and the resultant debris. Governments are addressing the issue through public education to help people recognize natural and official tsunami alerts and provide basic emergency instructions, improved warning systems, and effective evacuation planning.

Here are some guidelines for what to do before and during a tsunami:

1. If you feel a strong earthquake and you are close to water, immediately move inland to higher ground.

2. Do not wait and watch a tsunami come in. If you can see the wave, it may be too late for you to escape it.

3. Receding water is a warning sign that a tsunami may be coming. Get to higher ground immediately.

(Figure 19)

(Figure 20)

(Figure 21)

(Figure 22)

(Figure 23)

(Figures 21-23) Tsunami damage to the island of Okushiri in the M7.8 Hokkaido, Japan, earthquake in 1993. In figure 22, the remains of wood-frame houses cover the entire foreground, all destroyed and washed away by the roughly 30-foot tsunami. The only things left are the reinforced concrete foundations of the houses and small buildings and a few reinforced concrete structures, like the two-story one in the center. Figure 23 shows a close-up of remaining concrete foundations of houses annihilated by the tsunami.

TSUNAMI RISK IN THE UNITED STATES

The most recent destructive tsunami to hit the United States was generated by the great Alaska earthquake of 1964, magnitude 9.2. The peak wave height was 21 feet in Crescent City, California, where twenty-nine city blocks were inundated. It devastated many coastal settlements of Alaska—including Kodiak, Seward, Valdez, Whittier, and Cordova—causing a large proportion of the deaths associated with the quake. It also damaged settlements along the coasts of Washington and Oregon, killing several people and causing much property damage. Destruction resulted not only from the impact of the wave but also from the debris—logs, pieces of collapsed buildings, cars, fishing boats—carried by the water.

Two illustrations shown here graphically illustrate the power of the tsunami including the force of debris (in this case, able to pierce tires).

An earthquake on the Cascadia subduction zone would likely produce the United States' largest and most devastating tsunami. The Cascadia subduction zone is similar to the Sunda trench in Indonesia, which produced the magnitude 9.1 December 2004 Sumatra earthquake. Native American accounts of past Cascadia earthquakes, such as one in 1700, suggest tsunami wave heights on the order of 60 feet, which is comparable to those in the most damaged areas of Indonesia.

Tsunami inundation maps can be obtained for certain regions of the United States.

Alaska earthquake and tsunami, 1964

CHAPTER 3

Landslide Hazards

Landslides and rockslides occur frequently throughout the United States, causing a few billion dollars' worth of damages and more than twenty-five fatalities on average each year. In the West, the "landslide season" inevitably comes during the wet winter months, when rains saturate the ground, the water table rises dramatically, and "lubricated" hillside slopes begin to slide.

The landslide problem is particularly acute in the Los Angeles basin of Southern California, where thousands of large landslides have been mapped by the U.S. Geological Survey. This concentration is likely higher than in other comparably sized areas, but the San Francisco Bay Area and the Anchorage area follow closely, and the heavily populated areas of Utah are not far behind.

Earthquakes dramatically increase the potential for landslides. Major shocks trigger literally thousands of large and small landslides and rockslides throughout a stricken region. In the 1964 Alaska quake, extensive sections of the waterfront areas of Anchorage, Valdez, and Seward were destroyed by landslides. And in San Fernando in 1971, more than one thousand landslides, ranging in size from 50 to 1,000 feet, were triggered by faulting and the quake's vibrations. The 1989 Loma Prieta earthquake similarly triggered numerous landslides—particularly in the mountainous Santa Cruz area and along the coast from Monterey to Bolinas, north of San Francisco. Many homes were severely damaged. The Northridge earthquake in 1994 triggered thousands of landslides in the Santa Susanna Mountains. Fortunately, these California shocks occurred when the soil was reasonably dry. During a wet winter, many of the residential hillsides that were visibly ruptured by the quakes would have become landslides, and many more moderately shaken and damaged homes would have been totally demolished.

The Turnagain Heights slide in Anchorage during the 1964 quake was the largest and most spectacularly destructive single earthquake-generated landslide in a metropolitan area in the United States. During the tremor, a long bluff overlooking the sea broke into thousands of large earthen blocks and flowed outward toward the water, sweeping away a land mass nearly 2 miles long and 800 to 1,200 feet wide. The end of the slide extended far beyond the previous shoreline, and its head formed new bluffs as high as 50 feet. More than seventy buildings, among them some of the finest homes in the city, were carried 500 to 600 feet by the slide and were destroyed. Another slide, which occurred in

(Figure 24)

(Figure 25)

(Figure 26)

the middle of the business district along Fourth Avenue, dropped 11 feet vertically. All of the buildings in the area had to be removed, and most were complete losses. The city experienced several other smaller slides as well, causing tilting and broken foundations and other heavy structural damage.

For our purposes, there are three important points in the dramatic and tragic ground failure in Alaska. First, landslide hazards are eminently predictable. The geology of the Anchorage area had been investigated and recorded in great detail by the U.S. Geological Survey in 1959, and a published map indicated that landslides and slumps had occurred there in the past and would occur again.

Second, developers often ignore known landslide threats. Much of the area that slid in 1964 is again developed. Expensive homes in the Turnagain area and large buildings downtown have been built right over the old unstable terrain. The area that slid before will slide again—it is only a matter of time before another strong earthquake occurs.

Third, the geologic and seismic conditions that resulted in the Anchorage slides are found along innumerable bluffs and hillside "view" locations in Washington, California, Utah, and other areas of Alaska. But, as in Alaska, forewarning maps and reports are readily available for most of these areas.

The causes of landslides may be grouped into two general categories: natural geologic deficiencies and man-made problems. Natural geologic flaws are usually responsible for large slides, whereas smaller, localized slides that affect only one or a small cluster of buildings tend to be the result of development.

The most common natural condition for landsliding is a hillside, hilltop, or bluff in which the geologic foundation of rock or stable soil is layered by thin clay seams—sometimes so thin as to be virtually undetectable—that give way when they become saturated with water or the lateral forces of an earthquake break the clay's very weak frictional bond. The flow slide is a common form of natural landsliding, generally occurring in association with earthquakes. Flow slides are triggered by liquefaction under sloping ground. The liquefied soil simply flows away from its base in a muddy morass. The massive landslides in Anchorage were primarily of this type.

Developer-made slides occur on naturally stable slopes that have been disturbed by poorly engineered grading and landfill for homesites. Typically, a stable slope is cut and filled for a flat lot without providing support for the inclined soil layers at the uphill face of the cut. A great deal of rain and/or a moderate earthquake can loosen the freshly exposed inclined soil layers and send them sliding down upon the building below. Similar failures can occur with improperly graded downhill fills, in which water saturation or a moderate shock may cause the uncompacted landfill to slide, or the fill may overload the natural soil foundation at the bottom of the cut slope, causing failure of the whole hillside.

Many other factors are involved in the incidence of both natural and earthquake-caused landslides: the slope angle of the ground surface, the nature of the bedrock, the slope of the geologic layers, the existence of any surface or underground water flows, the level of the water table, the type and amount of vegetation that exists or is planted or removed, the proximity of active faults, the amount of rainfall, and so forth. Some landslide problems are correctable at a reasonable expense, particularly those that are man-made, which can be avoided in the first place by good engineering practices.

Several types of landslide restraints are possible, but generally their costs are excessive for most individual property owners. Also, landslide control by modern engineering techniques may become less feasible depending on the size of the potential slide area, the size and instability of any sliding that already has occurred, and the proximity of active faults. Finally, the success of a slide-stabilization program usually depends on constant surveillance and costly maintenance, and there can never be a full guarantee that sliding will not eventually occur anyway.

Keep in mind, though that if landslide risk is absent or can be eliminated, your hillside property represents one of the best possible investments in earthquake country.

(Figures 24 and 25) .The tremendous size and destructiveness of the 1964 Turnagain Heights landslide in Anchorage can be seen in these aerial and ground-level views of its aftermath. Studies published before the earthquake had indicated that the area was highly susceptible to slide damage, particularly during earthquakes.

(Figure 26) One of many landslides that damaged or destroyed expensive view homes throughout the Los Angeles area in the 1994 Northridge earthquake.

TYPICAL CAUSES OF LOCALIZED LANDSLIDING

Landslides in hillside subdivisions are not always the result of naturally unstable slopes. Frequently, the slides will occur where a basically stable soil foundation has been disturbed by careless and poorly engineered grading or landfill. Home buyers and builders should consult the available geologic maps, a geologist, or a geotechnical engineer to be sure that a hillside property has no past history of sliding and no apparent potential for failures. In addition, they should be aware of the characteristics of poor grading or fill and should investigate the stability of the slope above and below the house before committing themselves to the sale.

The building site shown in the top image has the natural ingredients of a landslide area: (1) The gradual rise of the natural slope is abruptly broken by a steep incline—an indication that a slide probably occurred at that point in the past. (2) A thin seam of clay lies between the bedrock and the upper soil levels. Only a careful geologic investigation—usually with drill holes—will reveal these thin clay seams, but many potential slide areas have been recorded and marked on the maps of various agencies and in the files of consulting engineers.

In the middle image, naturally stable ground has been disrupted by improperly graded or compacted landfill. New hillside developments are frequently stacked in a parallel series of cuts into the slope and landfills below the cuts to provide flat building sites. If the fill is not carefully engineered, it may settle or slide during an earthquake.

In the bottom image, there is no landfill problem. Instead, the cut slope above the house was graded too steeply, and loose soil along the cut tumbled down during an earthquake.

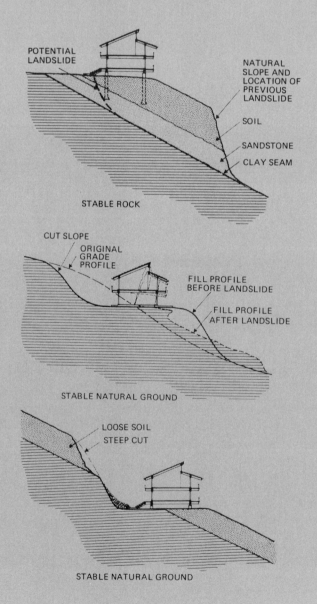

HOW TO IDENTIFY SIGNS OF LANDSLIDING

Cracks in the street above the house in the top left image indicate the head of the slide. Note that the asphalt is cracked in parallel fissures and the gutter is fractured.

The figure at the top right shows the weight of the sliding soil and house has buckled the sidewalk below the house.

In the middle figure, although it is not at all obvious to the eye, the house is leaning in the direction of the landslide. The tilt is visible only in the framing of doors and windows that open parallel to the slide. Here the door no longer fits well—a gap of about a half inch can be seen on the right side in the downhill corner. If several doors, windows, or cabinet doors show the same gap in the same direction, a homeowner should assume that the house is leaning and may be slowly sliding.

The lower right image shows typical cracks in the stucco of the same house. Such cracks tend to concentrate around openings (doors, windows, and so on) and in retaining walls.

Van Norman Dam, 1971

CHAPTER 4

Other Hazards to Your Property

Certain structures, such as dams and tall buildings near your property, can present special earthquake hazards.

Dams and Reservoirs

Dams and reservoirs present the greatest hazard to populated areas during an earthquake. The failure of a dam can ravage more buildings and claim far more lives than the shaking of the earthquake itself. Ironically, most major fault zones, such as the Hayward and San Andreas, the Newport-Inglewood, and the San Jacinto, provide the narrow canyons and natural sag ponds that are ideal for reservoirs, so that numerous dams serving the cities of the West have been built along or directly over recent scars of earthquake faulting. In addition, urban centers such as the Los Angeles Basin and the San Francisco Bay Area also contain hundreds of smaller reservoirs. The majority of these dams were built well before the principles of earthquake engineering were developed. Since the early 1970s, the State of California has been reviewing the safety of many of these dams. A number have been strengthened and some dismantled.

Several dams have collapsed in strong earthquakes. Certainly the most frightening recent earthquake-related dam failure in the United States was the partial collapse of the Lower Van Norman Dam in the Mission Hills overlooking the San Fernando Valley in the Los Angeles area. The massive 142-foot-high earthen dam, constructed in 1915, had been perfunctorily reinforced for earthquakes several decades later. However, the intense fifteen-second 1971 San Fernando tremor broke away almost its entire upstream surface, knocking the concrete facing and soil and rock into the reservoir. Fortunately, at the time, the reservoir was only slightly more than half full, and the slump of the dam still left a slim 5-foot margin of earth above the crest of the water. Had the water level been higher or the quake lasted an additional five seconds, or had a large aftershock struck before the reservoir could be lowered to a safe level, a flood of water would have swept down the hills and through a 12-square-mile area with some eighty thousand rudely awakened residents.

Other areas of California have not been so lucky. A similar failure of the earthen Sheffield Dam during the Santa Barbara earthquake of 1925 flooded the lower area of that city. A breach in the wall of the small Baldwin Hills Reservoir in December 1963, following several years of creep along the Newport-Inglewood fault zone in Los Angeles, claimed five lives and caused much property damage. Only the diligence of the reservoir's caretaker and the quick evacuation of the residential area below the reservoir prevented a much greater death toll. The Baldwin Hills Reservoir was a more modern structure, constructed in 1951, a time when the dangers of faulting were well understood. Yet it was built over the very fault zone responsible for the nearby destructive and carefully investigated Long Beach earthquake of 1933.

Even a carefully reinforced dam that is well removed from a fault can present an earthquake hazard to nearby residents. A sudden earthquake-induced landslide into a reservoir, for example, can damage the dam or turn the spill over the dam into a destructive flood. This type of event was graphically illustrated during the filling of the Vajont reservoir in the Italian Alps. After dinnertime on October 9, 1963, 7 billion cubic feet of earth plummeted into the lake, causing a 750-foot-high wave that crashed over the dam and onto the village below. Two thousand people lost their lives. The dam survived. A film released in 2001, *Vajont, La Diga del Disonore*, documents the tragedy.

Similarly, the sloshing of the contained water in a reservoir can generate waves that overflow the dam. Such waves, called seiches, topped the Hebgen Dam near West Yellowstone, Montana, during an earthquake there in 1959.

Three distinct waves overflowed the full length of the 720-foot dam and surged down into the unpopulated valley below.

What can you do about dam hazard in earthquake country? Since many dams, particularly in California, have been strengthened, you should first establish whether a nearby dam has been reinforced recently for earthquake loads. If it has, or if it is a new dam that meets recent California standards, the structure is probably safe. If the dam is old, was built pre-1972, and has not been evaluated and/or strengthened, you may face a very serious threat. Certain areas of the United States contain dam inundation maps.

(Figure 1)

(Figure 2)

(Figure 3)

(Figure 4)

Dikes and Levees

Typically, dikes and levees are built over and surround some of the worst possible geologic terrain for construction in earthquake country. The previous chapter noted that water-saturated alluvial or sandy soils along rivers and estuaries, landfilled areas over the mud of bays and marshlands, and water-saturated sand dunes along the seashore are subject to especially intense vibrations, settlement, landsliding, and liquefaction during an earthquake. Any of these effects can destroy a dike or levee, and thus any buildings that remain standing after the quake may be subjected to the additional and potentially worse damage of flooding.

The effect of failed levees on a metropolis was illustrated in 2005 when Hurricane Katrina slammed into New Orleans. The catastrophe was caused by unique (and known) vulnerabilities, and many dikes and levees are similarly at risk in earthquake regions throughout the United States. Perhaps the most vulnerable is that of the rapidly developing Sacramento–San Joaquin Delta in California, where approximately 1,000 miles of levees protect 700,000 acres of lowland. A major earthquake close to the delta would initiate extensive levee failures, causing widespread flooding of the Sacramento–San Joaquin Delta. Current economic impact estimates range from $10 to $30 billion. We knew that there would be a problem in New Orleans in a major hurricane, and we know that there will be a problem in the Delta in a major earthquake. We only hope that the Delta's levies are fixed before a quake comes.

Until the 1960s, little regulation or engineering analysis went into privately financed dikes and levees. Therefore, anyone living in marina developments or other low-lying waterside sites protected by such structures should investigate their engineering and earthquake resistance. A talk with a local civil or geotechnical engineer may reassure you that all possible measures have been taken to protect the dike or levee from earthquake damage.

(Figure 1) Lake Palmdale and smaller Una Lake are reservoirs created from the natural depressions, sag ponds, and scarps of the San Andreas fault zone near Palmdale in the desert of Southern California. The fault can be seen extending in a line from the dams through Leona Valley. Several smaller sag ponds lie in the distance along the trough of the fault. Sag ponds are formed when the fracturing and tilting of the ground from faulting brings underground water to the surface and then blocks drainage channels. Since sag ponds are generally enclosed on all sides, they provide ideal sites for reservoirs, but they are unfortunate and possibly dangerous locations for the dams that create these reservoirs. Lake Temescal in the Oakland Hills is another example of a reservoir formed by enlarging and damming a sag pond along a fault zone, in this case the Hayward fault.

(Figure 2) Creep along a branch of the Newport-Inglewood fault through Los Angeles caused the failure of the twelve-year-old Baldwin Hills Reservoir in 1963. Slow faulting under the dam slightly ruptured its concrete lining, and water then eroded a wider channel, which caused a large section of the dam to collapse. Several houses in the spill area outlined in the photograph were washed away by the flood, and many others were severely damaged.

(Figure 3) Even the largest of structures, such as massive concrete dams, cannot stop faulting. The Shihkang Dam was ripped apart by the fault that caused the M7.6 Central Taiwan earthquake of 1999. The roadbed on top of the dam on the right side of the photograph, where people are standing, was at the same elevation as the roadbed in the upper left corner of the photograph. It dropped 45 feet.

(Figure 4) The houses on the right side of Cloverdale Avenue in the Baldwin Hills area were completely washed away by the flood from the fault-creep-damaged Baldwin Hills Reservoir (see figure 2). The remnants of foundations and a swimming pool can be seen in the foreground.

(Figures 5 and 6) Broken levees led to massive flooding in New Orleans in the wake of Hurricane Katrina, a disaster that could repeat itself in the Sacramento-San Joaquin Delta (figure 6), where a fragile labyrinth of levees could be seriously damaged or fail (as the Sutter Bypass Levee failed in 1997) should an earthquake occur even as far as 100 miles away.

(Figure 5)

(Figure 6)

Semiattached or Taller Neighboring Buildings

Two semiattached or adjacent buildings with no gaps or only a small gap between their adjoining walls can seriously damage each other during an earthquake. Because the two buildings are structurally independent, they respond to vibrations in different ways and therefore pound against each other. This pounding can be especially severe at the roof level of the lower of two adjacent buildings.

In addition, a low building next to a taller building is threatened by debris falling from its neighbor. This is particularly true in larger cities, where a one- or two-story home may be separated by only a few feet from a much taller apartment or commercial building. If the taller building has unreinforced brick or concrete-block walls or veneer, precarious architectural features (such as chimneys, parapets, Spanish tile roofing, or poorly attached and heavy precast concrete panels), the lower building may be very seriously endangered. All of the deaths in Santa Cruz from the Loma Prieta earthquake of October 17, 1989, were caused by such debris.

If you own, rent, or use a structure adjacent to a taller structure, you may be at risk from falling debris. Consider the risks to your life and property, and consult a structural engineer for advice. You should also consider the purchase of earthquake insurance in such situations.

Row houses that are connected to each other do not suffer such damage because they move as a unit during a quake. Unattached row houses are also not usually vulnerable, particularly if they have plywood bracing in their ground floors (see chapter 6 for plywood bracing details). An exception is corner row houses or apartment buildings, which lack support on one side and are, therefore, free to deform and twist. Most all of the collapsed buildings in the Marina district of San Francisco after the 1989 earthquake were corner buildings.

(Figure 7)

(Figure 8)

(Figures 7 and 8) The pounding between these commercial buildings in Oakland during the 1989 Loma Prieta earthquake can be seen in the extensive damage to the facing and windows all along the adjoining walls of the two buildings. The same type of damage can be seen in the lower photo (figure 8) from Santa Cruz after the October 17, 1989, earthquake. The pounding from the lower building severely damaged the adjacent taller building and nearly collapsed it.
(Figure 9) A destroyed office building in Kobe in 1995, with one floor entirely collapsed. Had this building collapsed completely, smaller adjacent buildings would have suffered greatly.
(Figure 10) The collapse of an unreinforced and poorly connected brick wall caused this interior damage to a lower building in 1952.

(Figure 9)

(Figure 10)

(Figure 11)

(Figure 12)

(Figure 11) Inadequate separation between these two adjacent buildings in Mexico City probably led to the collapse of the taller one in the September 1985 earthquake. Note that the collapse occurred at the roof of the lower building, where maximum impacts were concentrated. Many structures, including at least one ten-story building in San Francisco, experienced similar damage but did not collapse in the 1989 earthquake.

(Figure 12) Older, taller buildings of which many can be found in our cities, represent high debris hazards to the smaller houses and apartment buildings below them.

Sichuan, China, 2008

CHAPTER 5

The Principles of Earthquake Resistance in Buildings

One could make the argument that earthquakes are no longer natural disasters; instead, they are human disasters. We know very well how to save lives and property—we just don't apply our knowledge sometimes. The massive death tolls associated with recent earthquakes in China in 2008, Palestine in 2005, and India in 2001, are not due to a lack of engineering knowledge. Nor, generally are they due to inadequate code requirements.

Structural engineers have made great strides over the past few decades in protecting life and property in earthquakes. Knowledge about the effects of earthquakes on buildings has steadily advanced during the years since the 1906 San Francisco earthquake, and the basic principles of earthquake-resistant design and construction have been well established since the 1950s. Inevitably, the chief constraints on the application of these principles to an adequate level in new buildings are cost and, as seen in India in 2001, a lack of process for ensuring code compliance in both engineering design and construction. In the United States, the greatest threat is not so much compliance with the building code, but aspects of the code itself. As described in several locations in this book, if the rules of the code are followed blindly, it is possible to design a building that meets the criteria of the code, but that may not meet the intent of the code. This is because the code is designed for certain types of structures under certain types of loading. If these rules are then applied to a different type of structure (e.g. a very tall building—see sidebar page 95), and the appropriate performance-based analysis and testing are not performed, the result may be a structure that will not perform as the code intended in a large earthquake.

Another problem is the consisten desire to lower costs of construction. Until all building owners, builders, and government agencies realize that a moderate increase in a building's construction costs will yield a substantially sturdier and more quake-resistant structure, earthquakes will continue to claim lives and inflict unnecessarily severe and expensive damage to property. Similarly, a modest outlay of money and effort on structural reinforcement can render most older buildings, particularly wood-frame structures, far safer both for their occupants and the investment of their owners.

The Effects of Earthquake Forces on a Building

The structural elements of any building are designed chiefly to distribute and then to carry the weight of the building and its contents and occupants to the supporting foundation and into the ground. This basic structural system involves some distribution of a building's weight along horizontal planes (beams, roof, and flooring, for example), but the heaviest load supported by the structure is along the vertical supports (columns and, particularly, walls) leading to the foundation. As we have learned, earthquake ground motions cause the ground to move horizontally and vertically in chaotic cyclic patterns. As the ground moves, so does the embedded foundation of a structure. The problem is that the structure, freestanding and flexible atop the foundation, does not want to move and resists with a force that is in proportion to its mass—its so-called inertial force. As the structure starts to move, it resists the motion of the earth as it cycles in another direction. As the ground, foundation, and structure move in this manner, forces and displacements develop in the building's structural elements and the connections between these elements.

Because buildings are, by their very nature, designed for large vertical loads, they generally resist the additional vertical forces of an earthquake effectively. However, horizontal earthquake forces can easily exceed the lateral strength of a conventionally built structure, and they usually result in damage to the elements of the building and foundation, and, all too often, the collapse of part or all of the building. Fortunately, these occur infrequently.

In earthquake country, special techniques for increasing the lateral resisting system in a building enable it to absorb and distribute the lateral forces of a tremor without much damage.

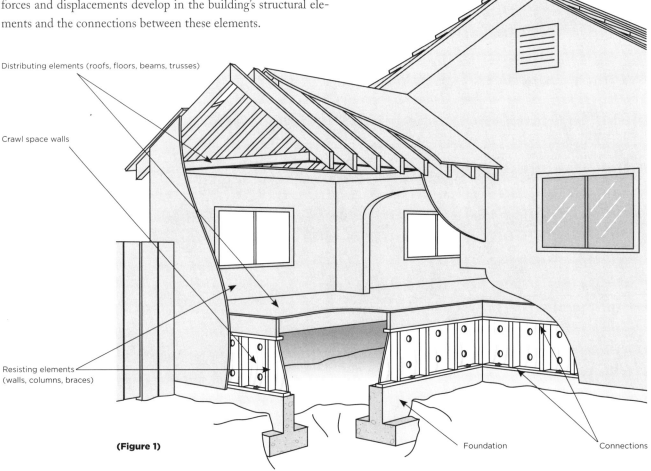

(Figure 1)

HOW A BUILDING RESPONDS TO GROUND SHAKING

During an earthquake, the ground waves cause lateral (horizontal) and vertical ground movements, or vibrations, which are transferred to a building through its foundation. The vertical earthquake movements cause the columns and walls of the building to contract and compress. This movement is usually not damaging, since buildings are, by their nature, designed to withstand large vertical loads. The lateral earthquake waves, however, are much more destructive because they are often the stronger waves, and horizontal strength is not the structure's prime purpose. As the ground moves, the structure will resist movement due to its inertia. The more mass, the greater the inertial force resisting this movement.

The effect of lateral earthquake movement on a building is shown in this drawing of a single-story structure. The movement emerges from the ground and travels through the foundation to the rest of the structure. The structure will naturally resist this movement, resulting in forces and deformations generated within the structure. The points of connectivity within the structure need to be specially designed to be able to withstand these force and deformation demands.

The earthquake waves inevitably focus on any weak connections or structural members, and once these begin to fail, the behavior of the building changes drastically. It is subjected to a chaotic mixture of new stresses and loads for which it is not designed, and the damage compounds until the building fails.

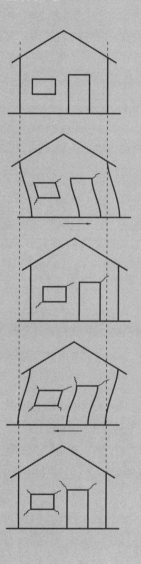

(Figure 1)
The basic structural components of any building:

    The *distributing structural elements* are those that lie in a horizontal plane. These diaphragms (roof and floors) and joists (beams and trusses) tie the walls together and disperse the static weight of furnishings, occupants and the elements themselves to the walls and foundation.

    The *resisting structural elements* are the vertical components of a building (walls, columns, and bracing). These elements support and transfer the load of the distributing elements to the foundation.

    The *foundation* supports and ties together the walls and transfers the weight of the building to the ground.

    The *connections* (nailing, blocking, joints, etc.) tie all of these components together.

Increasing Lateral Support

In an earthquake, the ground moves, and the building wants to stay in place. As a building resists being moved, forces are generated, and structural elements are stressed. The ability to successfully transmit inertia forces down to the foundation and into the ground while not compromising the gravity-load carrying capacity of the structure is the basis of earthquake engineering. The most common way to do this is through lateral bracing. Besides preventing building collapse, lateral bracing also limits damage because it reduces the displacements that the building will experience. Lateral bracing can follow three basic patterns: frame-action, shear-wall, and diagonal bracing.

EARTHQUAKE BRACING

Because of the severe lateral stresses to the walls and columns of a building during a strong earthquake, a special lateral bracing system is essential.

The frame-action bracing of steel or reinforced concrete is common in large buildings. Of the three types of bracing, it allows the most flexibility and movement, which can be a disadvantage—particularly to the occupants and furnishings of the swaying building. Shear-wall bracing is a solid, continuous wall of plywood (over a wood-frame) or reinforced concrete wall attached to the framing of the building. When this type of bracing is added as a supportive wall for the frame, it adds great strength to the vertical super-structure. Most important, shear-wall bracing is very stiff and unbending—thereby reducing the lateral deformations of the walls. This technique of bracing also provides a solid surface for the more direct transfer of the earthquake forces back to the foundation.

The use of diagonal bracing is another common bracing technique. Typically, a steel member is attached at an angle across the frame. Very often, two braces are attached in this way to form an X-brace. Both the diagonal and the X-bracing stiffen the framing of the building against deformations and provide a more direct path for the transfer of the earthquake forces to the foundation.

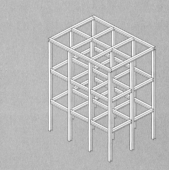
STEEL-FRAMED BUILDING

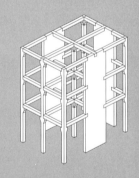
SHEAR WALL BUILDING

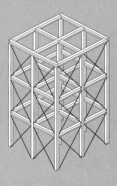
BRACED STEEL FRAME BUILDING

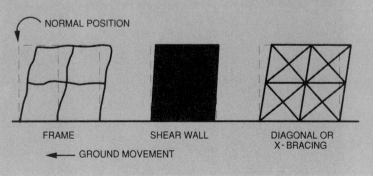

(Figure 2)

(Figure 3)

(Figure 4)

(Figures 2-4) Examples of retrofit construction using earthquake bracing for University of California-Berkeley buildings using frame action (figure 2), concrete shear walls (figure 3), and steel bracing (figure 4).

Frame-action bracing, used most often in large buildings, is composed of a series of connected frames of steel or steel-reinforced concrete that resist lateral earthquake forces by flexing. The bending action of the columns and beams of the frame absorbs the earthquake energy. Frame-action bracing is generally very effective in tall buildings, but it has one significant disadvantage: If the frame is too flexible, the bending action during a quake may cause large lateral deformations, leading to shattered or distorted exterior veneers, broken windows, plaster, and wallboard, falling ceilings, toppled furnishings, and pounding against adjacent buildings.

Shear-wall bracing entails the use of solid, continuous walls of plywood or steel-reinforced concrete. These walls add great lateral strength to the structure. Because shear-walls are relatively stiff and unbending, they also limit lateral deformations of the building and thus reduce the probability of architectural or interior damage. Wood-frame buildings can be made especially earthquake resistant by the addition of solid plywood sheathing to brace, tie together, and greatly strengthen their stud walls. The shear-wall bracing of large buildings is generally made of steel-reinforced concrete.

Like the other systems, diagonal bracing stiffens the supporting frame of a structure against damaging deformations during a tremor. It also increases a building's earthquake resistance by providing a more direct, diagonal path for the transfer of the structure's inertial forces to its foundation. Steel diagonal bracing is often used in steel-framed buildings and to strengthen older frame buildings. A single diagonal, two crossed diagonals (called X-bracing), or other patterns may be used.

The Building Materials

Certain building materials perform better than others under the duress of earthquake motion. Generally, wood and steel are preferred because (1) these materials are relatively light, which lessens the inertial forces that the structure must resist; and (2) these materials are tremendously flexible and can deflect and flex without cracking or breaking. (Too much flexibility can also be a disadvantage.) New concrete, concrete-block, and other masonry buildings can also be made safe, providing that special reinforcement procedures are carefully followed. Similarly, stucco over a wood frame without plywood sheathing requires special reinforcement details to prevent earthquake damage. These types of construction are described in the following chapter.

San Francisco, 1906

Kobe, Japan, 1995

CHAPTER 6

The Best and Worst Types of Construction for Earthquake Resistance

The basic structural materials of a building—wood, steel, concrete, masonry (stone, brick, or adobe), and the numerous combinations of these materials—are, with the lateral bracing system and other reinforcement measures, the most significant factor in the building's resistance to earthquake damage. Buildings have failed dramatically due to earthquakes throughout history, from 1906 in San Francisco to 1995 in Kobe, Japan, to 2008 in Sichuan, China. And buildings will continue to fail in subsequent earthquakes until builders and property owners pay adequate heed to earthquake engineering.

The more common forms of construction are discussed in this chapter.

Sichuan, China, 2008

(Figure 1)

(Figure 2)

Wood-Frame Buildings

A carefully designed and constructed modern wood-frame house or small building is the most desirable property investment in earthquake country. The high earthquake resistance of such buildings is primarily the result of the lightness and flexibility of wood. Lightness means that the load resulting from the building's inertial forces will be relatively small; flexibility enables the supporting components of the building—the wood-framed walls and columns—to bend during ground motion without breaking, or becoming disconnected. Similarly, since wood buildings are not so readily damaged by a tremor, they are better equipped to withstand aftershocks.

Wood-frame buildings are not by any means quake proof, of course. They accounted for 7,700 housing units rendered uninhabitable by the 1989 Loma Prieta earthquake and over 34,000 units rendered uninhabitable by the 1994 Northridge earthquake.

Engineers agree that a wood-frame building is most likely to suffer serious earthquake damage when one or more of the following conditions are present:

- It is built on unstable or soft ground.
- It is old (typically pre-1950).
- Its wood sills are not bolted down to its concrete foundation.
- Its crawl-space walls, or cripple walls, between the concrete foundation and the first floor are not properly braced with plywood.
- It has insufficient lateral bracing; for example, it has walls mostly of glass and practically no wood walls.
- It is built on slender vertical supports, such as stilts, used for some hillside houses.
- It has a heavy roof, such as clay tile.
- It has a "soft" ground story (in a building of two or more stories), with too many large openings such as garage doors and other big windows and doors on the ground floor.

Besides a sound geologic site and a strong, well-connected foundation, the most important element of a durable and safe wood-frame building is a lateral bracing system. Such earthquake bracing is now generally required for all new wood-frame construction in most earthquake-prone states. Some lateral bracing is also a common code requirement along the Gulf and Atlantic coasts, which are subject to the strong lateral forces of hurricanes.

Plywood sheets are by far the preferred material for wall bracing in houses. Like any other bracing system, shear-wall bracing with plywood panels will be effective only if its nailing is adequate. A variety of other products, such as Oriented Strand Board (OSB) and prefabricated engineered walls (such as Simpson Strong Walls), can be used in lieu of plywood. The conventional "double-wall" construction of sheetrock and stucco, very common in the western states, is simply not strong enough; such construction should be reinforced with additional shear walls of plywood. This is particularly important for buildings of two or more stories.

(Figure 1) As shown here, a carefully designed and constructed modern wood-frame house or small building is one of the best investments in earthquake country.
(Figure 2) Plywood sheathing—shear-wall bracing—when properly and adequately nailed is superior to any other type of bracing for a small wood-frame structure. Plywood shear-wall bracing is commonly found in earthquake country both as an exterior architectural feature (with battens, for example) and as a backing for stucco or some other exterior facing.

Wood diagonal bracing is outdated for earthquake country. This entailed the mounting of a series of lumber strips or metal straps at an angle across the studs of a building. Such diagonal bracing is not nearly as strong as shear walls of plywood.

The traditional wood siding of the majority of older houses and many modern structures is much weaker than shear-wall bracing with plywood. But if the siding is of high-quality lumber, and if the boards are fitted tightly together and are well nailed and maintained, the walls should survive a moderate quake. Wood-shingle siding is much weaker and does not provide the strength of shear-wall bracing with plywood.

Some older (typically pre-1950) houses are completely sheathed with diagonal boards. This diagonal sheathing is traditionally done with 1-by-6-foot boards, and it is very strong if properly nailed. Any wood rot of the diagonals, particularly near the sill, or of the sill itself drastically reduces the effectiveness of the sheathing, however.

(Figure 3)

(Figure 5)

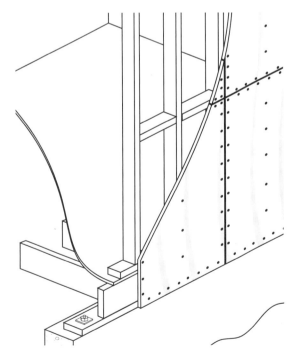

(Figure 4)

(Figure 6)

(Figure 7)

(Figure 8)

(Figure 3) Wood (or metal-strap) diagonal bracing, shown here, is not recommended. Instead, use plywood sheathing. Diagonal bracing in houses has shown itself to be inadequate in many earthquakes, often leading to severe damage.

(Figure 4) This drawing illustrates the appropriate means of connecting plywood sheathing. Note the frequent use of nailing.

(Figure 5) An example of a well-braced wall.

(Figures 6 and 7) Good plywood bracing details are shown here. Note the plywood panels next to the windows. Figure 7 shows dense nailing around the periphery of the plywood panels.

(Figure 8) The lower floor of this two-story house is completely sheathed with plywood.

Wood-Frame Buildings with Stuccoed Walls

Although stucco is an easily fractured material, it is not as fragile during earthquakes as one might think. If stucco is applied to strong wire mesh that has been carefully lapped and securely nailed to plywood sheathing (or to the clapboard of older frame buildings), it will seldom fracture or fall in even a large earthquake, because it is quite flexible and strong. Many stuccoed buildings, unfortunately, are constructed with only diagonal bracing and sheetrock or plasterboard backing, and neither is strong enough to withstand earthquakes without significant deformations that crack and break the stucco. Multistory stucco buildings without shear-wall plywood bracing are especially susceptible to such damage because the greater inertia of their upper floors causes greater deflections.

(Figure 9)

(Figure 10)

(Figure 9) A house under construction in California in the 1970s. A few days after the photograph was taken, the house was stuccoed. No plywood was used to back up the stucco and to provide the additional stiffness required for good performance in a strong earthquake. If any cracks occur in the stucco, which is a near certainty, then the brace, with its very few nails, will provide the only resistance on the ground floor.
(Figure 10) Multistory stucco buildings, in particular, are subject to extensive damage. Note the stucco damage to an apartment building in the San Fernando Valley.
(Figure 11) The wall cross-section (left) shows the recommended procedure for applying stucco to metal lath backed by plywood sheathing. The detail (right) illustrates a procedure recommended for strengthening existing stucco walls. Wood sheathing stiffens the walls and lessens deformations that could break the stucco.

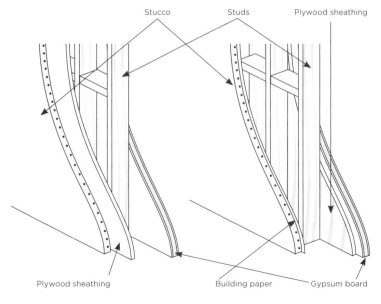
(Figure 11)

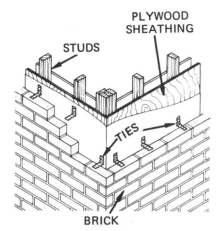

(Figure 12)

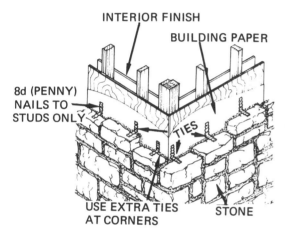

(Figure 13)

Wood-Frame Buildings with Masonry Veneer

Masonry veneer is any brick, stone, or other such covering—usually on outside walls—that is attached to the main wood framing. You can tell whether brick is veneer by checking whether the inside walls are brick. If they sound hollow, as interior walls do, or if you can drive a nail easily through the Sheetrock or plaster, you probably have veneer. Most often, looking into the crawl space or basement is sufficient; you should be able to see the wood framing if the exterior is veneer.

Masonry veneer does not support any load from the house, and, though it can be an appealing architectural feature, it is also highly susceptible to earthquake damage. The additional weight of a masonry façade generates greater inertial forces that can cause serious structural damage as well as damage to the veneer itself. Further, as the anchorages between the brick or stone and the wood frame are often weak or insufficient, the veneer can crack or be torn away from the frame during a tremor.

Multistory veneered buildings, especially, often suffer extensive and expensive damage to their masonry coverings. Most earthquake insurance policies reflect this fact with the inclusion of addenda excluding masonry veneer from coverage.

(Figures 12 and 13) Reinforcement for masonry veneer: These views of masonry-veneer construction show the proper placement of anchors.
(Figure 14) This apartment building in the Marina district of San Francisco lost much of its poorly attached brick veneer during the 1989 Loma Prieta earthquake.

(Figure 14)

CHAPTER 6

(Figure 15)

Low-quality mortar is another factor in damage to masonry veneers. Lime mortar is particularly notorious: it has no resistance when subjected to earthquake forces. If you own or wish to purchase a veneered building, examine the hardness of the mortar. Scrape a coin across the mortar, and if it falls away easily, it is very likely to fare poorly during an earthquake.

Because stone veneer is usually heavier than brick and is placed in irregular patterns, its anchors must be very carefully placed. It is not advisable to cover large areas, such as entire walls, with stone veneer—particularly when the building is in an area that may experience high earthquake intensities. The very probable failure of the heavy veneer can seriously damage the frame of the building and could be a hazard to nearby property, to passersby, and to the occupants.

(**Figure 15**) This Coalinga, California, house was reinforced with plywood shear walls, which prevented any serious structural damage in the 1983 quake, but its concrete slump block veneer was not adequately tied to the plywood. Several brand-new houses suffered similar damage in an earthquake with a magnitude of less than 5.0 in May 1990 in Alamo, along Northern California's Concord and Calaveras faults.

(**Figure 16**) These older wood-frame houses in San Fernando were stripped of their stone veneer by the 1971 tremor. It is probably fortunate that the heavy veneers were not anchored to the frames, for the collapse of the stones might have taken large portions of the structures with them. Note that the glass is not broken.

(Figure 16)

Concrete-Block Buildings

When the hollows of concrete blocks are properly reinforced with vertical and horizontal steel rods and then carefully grouted with poured concrete, building walls made of such blocks form solid and continuous shear-wall units. Reinforced concrete-block buildings can exhibit great strength and resistance under the stress of earthquake forces. However, because the walls are an assemblage of separate block units joined by concrete and mortar, they can share many of the weak points of brick construction if the concrete does not completely fill the cavities, if their steel reinforcing is inadequate, if their connections with the building's diaphragms are weak or insufficient, or if the engineering design is poor.

Any of these mistakes in structural design or workmanship can cancel the advantages of concrete-block construction. For example, a large percentage of the damage, injuries, and more than five thousand fatalities in the 1972 Managua, Nicaragua, earthquake was caused by the partial or complete collapse of poorly designed and constructed concrete-block buildings. All unreinforced concrete-block structures are extremely dangerous in even moderate earthquakes.

(Figure 17)

(Figure 18)

(Figure 19)

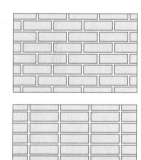
(Figure 20)

(Figures 17 and 18) In both of these Alaskan structures, the concrete-block walls were unreinforced and therefore collapsed. The results of such inadequate design and construction are evident here. Neither the Valdez hotel (figure 17) nor the Anchorage apartment house (figure 18) were salvageable after the 1964 quake.
Note the undamaged wood-frame structures on both sides of the hotel.
(Figure 19) This severe quake damage to a concrete-block structure illustrates what happens when the blocks contain no grouting, which makes the reinforcing steel completely ineffective in strengthening the wall.
(Figure 20) The two most common patterns for concrete-block construction are the staggered common bond (top) and the vertically aligned stack bond (bottom). The former is recommended for earthquake country because of its greater strength and resistance to lateral earthquake forces.
(Figure 21) The stack-bond, concrete-block walls of this San Fernando building were reinforced with vertical steel bars but lacked sufficient horizontal bars and concrete grouting. A staggered common-bond pattern might have lessened this quake damage.

(Figure 21)

Reinforced Masonry Buildings

Reinforced masonry buildings are very similar in construction to concrete-block buildings. Two separate layers of brick are laid with connecting steel ties embedded in the mortar, and horizontal and vertical steel reinforcements are then inserted in the space between the two layers. The steel rods are tied together, and the space is then filled with poured concrete. Properly constructed buildings of this type have proved to be effective in resisting at least moderate earthquake forces and are acceptable for earthquake country.

Clay-Tile Buildings

Hollow clay tile is similar to concrete blocks, save for one important difference—clay tiles are very brittle and easily shattered. Therefore, clay tile is simply not a sound building material for earthquake areas. The best course is to avoid such buildings.

(Figure 22)

(Figure 23)

(Figure 22) Because of their extreme brittleness and inflexibility, hollow tile structures, such as this guest house of the Veteran's Hospital complex in San Fernando, inevitably suffer proportionally greater earthquake damage than most other masonry buildings.
(Figure 23) Many old (typically pre-1940) office and other commercial buildings have partition walls built of hollow clay tile blocks. Such walls break apart easily and create hazards, as illustrated by this damage to the fire and emergency stairwells of a department store building in Oakland after the 1989 Loma Prieta earthquake.

Residential Steel-Framed Construction

A well-designed and built steel-framed house can be even more earthquake resistant than those built of wood. Steel has been used in some residential construction for many years, and starting in about 1980 it became more common. By the mid-1990s, hundreds of builders were using steel in residential construction, and, as steel studs and other steel-framing systems have become more readily available, the growth has continued.

There are various types of steel-framing systems, from simple stud construction to wall panels to pre-engineered steel channels. In all of these systems, the components are either screwed, bolted, or welded together, providing a connection that is more positive than that achieved with the use of nails. This is a major advantage in earthquake country, as steel framing provides consistent strength and secure connections throughout a structure.

(Figure 24)

(Figure 25)

Steel-Framed Buildings

Typically, these are the safest commercial buildings.

Most high-rise buildings in California are built around a steel frame (although very tall buildings are increasingly being constructed with reinforced concrete only without any significant steel framing—see sidebar on page 95). Steel buildings tend to be relatively earthquake resistant due to their light frame and ability to absorb a great amount of energy and deformation. This ability to deform, however, was called into question when over two hundred steel buildings experienced fractured critical connections during the 1994 Northridge earthquake. Similar problems were discovered after the 1995 Kobe earthquake. Factors contributing to these failures—connection details, welding procedures, and lack of quality control during the welding process—have largely been addressed in new code provisions since these two earthquakes.

The safest steel-framed buildings are those constructed after 1997, although structures built after 1973 are much safer than buildings built before this date (mainly due to U.S. code revisions following the 1971 San Fernando earthquake).

The main disadvantage of steel-framed structures is in their flexibility. If the structure is too flexible, nonstructural elements such as plaster walls, partitions, ceilings, and facades often crack or distort. Older (and some newer) steel structures with heavy, ornate exterior facades (mostly of unreinforced brick, stone, and terra-cotta) have suffered expensive damage in previous earthquakes. Unreinforced interior partition walls, too, often crack and sometimes shatter.

The addition of concrete shear walls, particularly around the elevator cores of large buildings, provides all of the advantages of a steel building while reducing the flexibility of the typical steel-framed building. This is called a Dual System, and it is probably the most reliable of all structural systems (for medium rise structures) for strong earthquakes. The system also provides redundancy—if the steel framing is damaged, the concrete walls can carry the additional loads, and vice versa.

(Figures 24 and 25) Steel-frame houses. The top photo (figure 24) is a view of the finished steel framing and the lower photo (figure 25) shows the finished house. Steel-frame houses are a good investment in earthquake country.

(Figure 26)

(Figure 27)

(Figure 28)

(Figure 29)

(Figures 26 and 27) Old steel framed buildings with facades or partitions of unreinforced masonry, including brick and terra-cotta, often suffered very costly damage in the 1989 Loma Prieta earthquake. Note the steel framing (figure 27) in this 1920s Oakland building.

(Figures 28 and 29) One of the important overall surprises of the Northridge earthquake of January 17, 1994, was the widespread and unanticipated fractures in welded steel beam to column connections. No casualties or complete collapses occurred during the Northridge earthquake as a result of these connection failures. Another ten seconds of shaking, however, and this might have been a different story.

Figure 28 shows research into new improved steel beam-column connections. The "damage" is directed to the beam (the horizontal member), as opposed to the columns (the vertical member), the gravity supporting elements (if the columns fail, the whole building falls down, but if the beams fail, you usually only get localized damage, which is not nearly as catastrophic). New buildings are built with these concepts in place, but many older steel frame structures still need retrofitting.

Figure 29 shows a form of beam-column connection called the "dog-bone" in which structural damage is concentrated in the tapered section of the steel member.

(Figure 30)

(Figure 31)

(Figure 32)

(Figure 30) The steel frame of the San Francisco–Oakland Bay Bridge collapsed during the 1989 earthquake at one location because of inadequate connections between the beams supporting the roadway and their supports.

(Figure 31) Architectural damage in the M7.4 Sendai (north of Tokyo), Japan, earthquake of 1978. Unnecessarily heavy concrete panels were not properly connected to the steel frame; they caused damage to the steel structure behind them.

(Figure 32) Steel buildings, like this one in Spitak, Soviet Armenia, were practically the only structures still standing in the destroyed city of 40,000 people. This industrial building lost its poorly attached stone and precast concrete façade but remained standing, practically undamaged.

Concrete Shear-Wall Buildings

Like steel buildings, reinforced concrete buildings that utilize shear walls and were properly designed have a good track record in strong quakes. They are usually good investments and relatively safe places to live or work.

Concrete shear-wall buildings rely upon massive, stiff concrete walls, rather than frames supported by concrete columns, for their structural integrity. The walls are typically built together with a concrete frame. The performance of concrete shear-wall buildings is highly dependent on the number of walls, their location within the building, their configuration, the size and number of openings in the walls, and reinforcement details.

The failure of the Olive View Hospital in Los Angeles during the 1971 San Fernando Valley quake is an instructive example into the use of shear wall construction; it was also a major turning point in earthquake engineering. The hospital had been completed and dedicated only a few weeks before the earthquake, and it was presumably designed and constructed in accordance with the latest building codes. It collapsed because concrete shear walls were eliminated on the ground floor to allow for more window space (soft-story condition). In addition, the slim concrete supporting columns at street level were inadequately reinforced with steel, and the combination of the heavy inertial load and the chaotic deflections of the ground floor shattered the columns and caused the entire structure to lurch to one side. Three of the four independent exterior multistory stairwell structures totally collapsed because of the inadequate columns. Several new adjacent satellite buildings also collapsed or were severely damaged. Many high-rise concrete buildings designed to standards similar to those of California prior to 1973 have also failed or completely pancaked in recent earthquakes in the state, as well as in most major earthquakes worldwide, and particularly in the Kobe, Japan, earthquake of 1995, and the Taiwan earthquake of 1999.

Irregularly shaped shear-wall buildings and those with a soft story often perform poorly in earthquakes, mainly due to torsion (twisting). In addition, buildings with walls distributed around only two or three sides are vulnerable to strong twisting forces and have been severely damaged in past earthquakes.

(Figure 33)

(Figure 35)

(Figure 34)

(Figures 33–35) The reinforced-concrete Olive View Hospital in San Fernando was dedicated only weeks before the 1971 earthquake destroyed it, causing four deaths and several injuries. Three of its four separate stairwell structures fell to the ground and two smaller concrete buildings in the complex suffered partial collapses. Such failures in new, presumably well-designed structures during a moderate quake demonstrated weaknesses in the engineering application of contemporary building codes. Much has changed since then, mostly for the better.

THE RISK OF NEW TALL CONCRETE BUILDINGS

One of the developments of the real-estate bubble of the first decade of the 21st century is the construction of tall reinforced concrete buildings in regions of high earthquake hazard. Until recently, almost all tall buildings in earthquake regions of the United States were built of structural steel. Most of these were also commercial buildings.

Tall reinforced concrete buildings are now the norm for residential construction such as condominium towers. Many are slender, seemingly all glass buildings with panoramic views uninterrupted by structural elements such as columns and shear walls. Some of these new residential towers rely on a substantial concrete shear-wall core to provide earthquake resistance. In effect, these buildings resemble in their overall design a tall industrial concrete stack or chimney, surrounded by columns and enclosed by glass.

There are three reasons for concern regarding the earthquake safety of these buildings. The first concern is that U.S. building code requirements for earthquakes were not developed and written for application to tall buildings (say over 20 stories). Prescriptive rules in the code are based around the dynamic behavior of low- and medium-rise structures, not the more complex dynamic behavior of tall buildings. If the code procedures are followed blindly, and low- to medium-rise building analysis and design procedures are used for high-rise construction, the result can be a building that is more vulnerable than that anticipated by the codes. It is essential that tall buildings be designed by highly qualified structural engineers (as many of them are), and that the appropriate earthquake hazard assessment, structural analysis, testing, and peer review be performed. This concern has recently been recognized by the regulatory authorities in San Francisco and Los Angeles, for example, where the need for advanced methodologies for designing high-rise buildings is now recognized.

The second issue is the potential lack of redundancy in the structural system. The same structural elements that are intended to absorb the earthquake energy (the shear walls) are also primary gravity-supporting elements. If the shear wall fails due to unexpected loading or unanticipated structural behavior, there is no second line of defense to prevent the tower from collapsing.

The third issue is our lack of experience with the behavior of these tall concrete buildings in very long and large earthquakes. Many capable engineers and engineering professors have spent much time analyzing the buildings, and tests have been conducted at universities and elsewhere to understand their behavior. However, these studies are ongoing and unanswered questions remain. As none of our modern construction of any type has been tested under very strong ground motion (we have yet to experience a great earthquake in the mainland U.S. in modern times), we do not have the evidence to demonstrate the ultimate effectiveness of any new design and construction technique. After the authors' visits to - and investigations of - more than 100 earthquakes around the world, however, we have learned that earthquakes do cause unexpected events do happen.

The upshot is that there are potential risks associated with tall concrete buildings. Their performance and safety is building-specific, they sometimes lack redundancy, and they embody unanswered questions and differences of opinion within the international earthquake engineering community. This is all underscored by the large numbers of people living and working in these buildings.

If you are considering investing, working, or especially living in one of these buildings, we recommend that you first do your research, including seeking the advice of a couple of highly qualified structural engineers who are familiar with the design of tall buildings in California or Japan.

Note the differences in the performance of the two types of columns supporting the structures. The completely fractured corner column in the foreground of figure 34 had vertical steel reinforcements that were tied together with individual steel loops. These loops pulled apart, allowing the concrete core to burst. The other columns were reinforced with closely spaced and continuous spiral reinforcements that held the concrete columns together.

In the 1980s, the hospital was torn down and completely rebuilt with a massive steel frame (figure 35). The replacement hospital is shown just after the 1994 Northridge earthquake, which was as strong as or stronger than the 1971 earthquake. The replacement hospital was designed to California's new earthquake hospital code and suffered no structural damage.

Concrete-Frame Buildings

A modern, properly designed concrete-frame building can be earthquake resistant, although this is not the case for many, if not most, older structures. So-called nonductile concrete-frame structures exist in the San Francisco and Los Angeles areas, Seattle, and St. Louis by the hundreds, including dozens of high-rises. A number of these buildings were damaged in the 1989 San Francisco earthquake to the point of near collapse. In fact, this was the type of construction used in the Interstate 880 double-decker Cypress structure that collapsed in Oakland, killing forty-two people. In the Northridge earthquake, the most dramatic building collapses were those of the older concrete-frame buildings, including the Kaiser Medical Building, many buildings in the Northridge Fashion Mall, and the parking garage at Cal State Northridge. Fortunately, the earthquake struck early in the morning, when the collapsed buildings were mostly unoccupied.

Concrete-frame structures generally use concrete beams and columns in somewhat the same manner that steel-frame buildings use steel beams and columns. Designing adequate connections and reinforcing steel detailing for these much heavier

(Figure 36) Older, nonductile concrete frame structures are the most hazardous large structures in earthquake country. The 1950s I-880 viaduct in Oakland, California, lacked structural details required by the current code. Several office buildings of this type of construction in San Francisco, Oakland, and elsewhere came to within seconds of collapse on October 17, 1989. Some of the severely damaged buildings were designed and constructed as recently as 1981.

(Figure 37) Because of their more unusual architectural configurations, such as large and open reception areas, ballrooms, etc., hotels are typically some of the highest risks in earthquake country. This concrete frame (without shear wall) hotel in the resort city of Baghio was demolished by the M7.7 1990 Central Luzon, Philippines, earthquake.

(Figure 36)

(Figure 37)

(Figure 38)

(Figure 40)

(Figure 41)

(Figure 39)

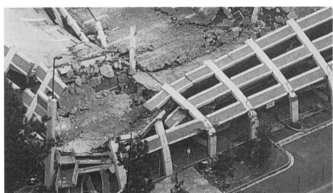

(Figure 42)

(Figures 38 and 39) Nonductile detailing was largely to blame for the collapse of this large section of the Hanshin Expressway during the 1995 Kobe, Japan, earthquake (figure 38). The details that caused the failure of the columns are similar to those that caused the failures of the older freeways in the Los Angeles area in the 1994 Northridge earthquake (figure 39).

(Figures 40–43) Failed precast concrete buildings in the Northridge earthquake in the San Fernando Valley. These are all structures that lacked adequate earthquake resistance despite being relatively new. Figures 40 and 41 depict a damaged garage at the Northridge Shopping Center; figure 42 shows a structure at California State University at Northridge; figure 43 shows a medical office building. A few shear walls and better connections would have prevented these collapses.

(Figure 43)

(Figure 44)

(Figure 45)

(Figure 46)

(Figures 44 and 45) Two examples of collapsed multistory commercial nonductile concrete frame buildings in Mexico City in 1985. Many comparably weak, older (typically pre-1973 buildings in California) concrete buildings can be found in San Francisco, Los Angeles, Seattle, Utah, and elsewhere.

(Figure 46) This reinforced concrete frame school building had been strengthened with new concrete shear walls just before the M6.8 Niigata earthquake of 2007. Two of the solid walls can be seen in the middle and near the right end of the structure. It was necessary to block some of the classroom windows in order to insert the new walls. The school suffered no damage, demonstrating once again that older concrete buildings can be strengthened at a reasonable cost.

(Figure 47)

(Figure 48)

(Figure 49)

(Figure 50)

(Figure 51)

(Figure 52)

(Figures 47 and 48) Collapsed newer reinforced concrete buildings in the M7.6 Taiwan earthquake in 1999.

(Figures 49–52) Several shopping malls, including the Bullock's department store in the Northridge Shopping Center (figure 49), were damaged by the 1994 M6.7 Northridge, Los Angeles, earthquake. Figure 50 shows part of the interior of the store. Note that the second-and third-floor slabs have collapsed through the columns and are now pancaked onto the ground floor area. Figure 51 shows another department store that has collapsed partially (see the sagging center of the roof). One of the strengthened stores is shown in figure 52. Built in the 1930s and strengthened a couple of years before the earthquake, it suffered no damage.

building elements is, however, a far more difficult engineering and construction feat.

In the magnitude 6.5 San Fernando Valley quake of 1971 and again in the similar 1994 Northridge quake (whose epicenters were just a few miles apart), many concrete-frame structures in high-intensity areas suffered severe damage or collapsed. Engineers found several common design flaws, among them insufficient ties holding the vertical reinforcing steel together and confining the concrete in columns, and poor reinforcing of beam-to-column joints. As a result, code requirements for these details were extensively modified in the 1973 Uniform Building Code (UBC), which governed earthquake design in the western United States until the recent introduction of the International Building Code (IBC). Further refinements were made in later editions of the codes. Consequently, concrete-frame structures designed after these code revisions took effect should perform substantially better than those designed earlier.

One of the most significant design problems for concrete-frame structures—and indeed for any structure—is a ground floor with large openings for garages or windows. Although aesthetically desirable, these so-called soft-story designs (see next chapter) leave little room for bracing, thus weakening the structure's ability to withstand an earthquake. Building codes after 1988 have restricted soft-story configurations and require special design considerations when they are used.

Many dangerous larger older buildings exist throughout earthquake country. This particularly includes many apartment buildings, especially high-rise apartment buildings.

If you live or work in a pre-1973 unretrofitted concrete-frame structure (especially one with an open first level), you are exposing yourself to one of the greatest hazards in earthquake country. Overall, these buildings are the worst earthquake hazards in the United States, primarily because so many thousands have not yet been strengthened. This applies to multistory commercial and apartment buildings. If you live in one of these buildings, you

WHY OLD BUILDINGS ARE GENERALLY MORE DANGEROUS

There is much debate about the quality of earthquake construction after the 1868 and 1906 earthquakes in the San Francisco Bay Area; although new research demonstrates that architects and engineers were building with earthquake resistance in mind well before the nation's first mandatory earthquake code came into effect following the devastating 1933 Long Beach, California, earthquake. Regardless, it was not until after the 1971 San Fernando earthquake that we took a step change in designing buildings for earthquake safety. Changes were reflected in the 1973 Uniform Building Code.

Unfortunately many nonretrofitted buildings built before 1973 are dangerous. The good news is that many buildings have been retrofitted; the bad news is that more have not. In particular, there are many vulnerable midrise concrete apartment buildings throughout earthquake country. The skylines of San Francisco (right), and Los Angeles are peppered with dangerous pre-1973 buildings.

If you live or work in a nonretrofitted building that was constructed before 1973, especially a concrete building, you should seriously consider moving or applying pressure to your company and/or landlord to have the structure assessed and upgraded.

San Francisco, California

could lobby your landlord to upgrade the building, but perhaps the easiest solution is to move.

Concrete Tilt-Up Buildings

The concrete tilt-up building is what we see in most modern U.S. industrial parks. Much of the Bay Area's Silicon Valley, Orange County's high-tech industry, the Puget Sound area, and Salt Lake City's industrial parks are housed in such structures, for they are cheap, easy, and fast to build.

Concrete tilt-up design came into general use in the early 1950s; today it is one of the least costly industrial and commercial structures to construct. The concrete foundations and base slab are poured in place; next, the walls are poured in place, lying horizontally on the ground, and are then tilted up, like cards, on top of the foundation. Usually, a wood roof is then built, with only a few slender interior columns supporting it.

These structures can be extremely dangerous, especially if built prior to 1973. This often makes for a risky investment and a potentially dangerous place to work. The tilt-ups' primary structural vulnerability is their roof-to-wall connections, which can break and allow the concrete walls to separate from the roof and to fall—or tilt back—to the ground. Unlike steel structures that are able to deform without failure, once a tilt-up structure

(Figure 53)

(Figure 54)

(Figure 55)

(Figure 53) A collapsed tilt-up after the Northridge earthquake. The structure failed at the roof-to-wall connection.
(Figure 54) This tilt-up was strengthened at a very reasonable cost just before the 1994 earthquake. Unlike its neighbor (figure 53), it survived without significant damage.
(Figure 55) This concrete tilt-up, light-industrial building in San Fernando was one of the many that were badly damaged by the 1971 quake. Again, in almost every case, the roof-to-wall connections were poorly designed—a very common weakness in larger concrete and masonry buildings with wood roofs and floors.

(and especially its connections) reach its design force limit, it will likely collapse, with disastrous consequences.

Following tilt-ups' poor performance during the San Fernando Valley quake, relevant seismic requirements of the UBC were modified in 1973. The UBC was again revised in 1976, with more stringent requirements. Since then, and after pretty much every major earthquake, the code has been further altered to try to improve the performance of tilt-ups during strong earthquakes. Tilt-up buildings designed to these new requirements should generally perform much better than older structures, although they may still suffer significant damage and be unusable following an earthquake. The ironic fact remains that much of the United States' high-tech industry is run in these low-tech (and much-too-often-dangerous) structures. Even apart from life-safety concerns, the chance of major business interruptions—in even moderate events—is very high.

The best news about such buildings is that they typically can be strengthened and fixed for a very reasonable price. In fact, in the past few years, numerous California companies undertook extensive programs to evaluate their tilt-ups and fix those found to be dangerous. Some tilt-ups have been strengthened outside California, too, but other earthquake-prone states—Washington, for example—lag badly and sadly behind.

Typically, tilt-up buildings that had been retrofitted prior to the 1989 Loma Prieta and 1994 Northridge earthquakes performed very well in those earthquakes. In a number of cases, they had no damage, whereas nearby, nonretrofitted or unstrengthened contemporary tilt-up buildings fell apart.

(Figure 56)

(Figure 57)

(Figure 58)

(Figures 56 and 57) New tilt-up buildings under construction. Note that all of the strength of the buildings is in the exterior walls; the interior frame supports the roof only.
(Figure 58) An example of damage to tilt-up buildings in the Northridge earthquake. The damage could easily have been prevented with the addition of a few bolts between the collapsed wall and the roofing beams.

(Figure 59)

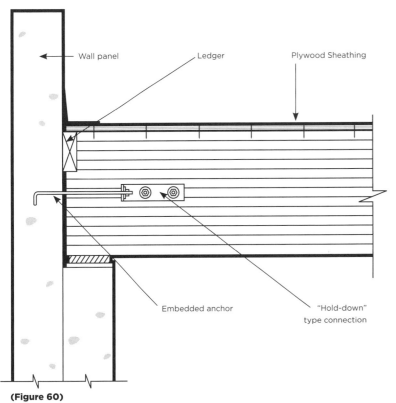

(Figure 60)

(Figures 59 and 60) A collapsed tilt-up building in Hollister after the 1989 earthquake (figure 59). Stored pallets of tomato products fell against the poorly connected exterior walls and pushed them out. The addition of a few "hold-down" connectors, such as that illustrated in figure 60, would have prevented the damage.

CHAPTER 6

Unreinforced Masonry Buildings

Unreinforced masonry (URM) buildings (including houses) have long been recognized as some of the most hazardous smaller structures in earthquake country. The major flaw of these structures is that they are brittle and cannot deform without being damaged by the lateral thrusts of an earthquake. Their brick is heavy and inflexible, so lateral motions create an overwhelming inertial load that cracks the usually weak mortar connections (the glue that holds individual bricks together) and causes the bricks to separate. Once this cracking occurs, the entire building can collapse progressively.

The quality of the mortar is particularly important to the performance of such buildings. When it is poor or old (and usually it is both), lateral earthquake stresses form cracks in a zig-zag course through the mortar and around the bricks. During large shocks of long duration, such cracks tend to propagate diagonally across the walls and rapidly reduce their strength and continuity. In California, mortar in old buildings is so bad that it has earned the nickname "buttermilk mortar."

To date, URMs have caused the majority of earthquake deaths in the United States. A study after the Bakersfield quake of 1952 found that only one of the seventy-one older brick buildings in the city had survived the tremor undamaged, and more than thirty buildings had to be either torn down or substantially revamped by removing one or more damaged upper stories. Coalinga, unlike most other earthquake-prone U.S. cities, no longer has a URM problem because after the 1983 earthquake there, most of the ninety brick buildings downtown were removed—they had either collapsed or were too severely damaged to be fixed. In the 1989 Loma Prieta earthquake, most deaths, aside from those at the collapsed Cypress freeway structure, were caused

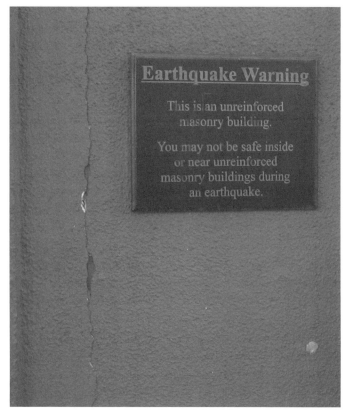

(Figure 61)

(Figure 61) In some cities, it is required that all URM buildings have warning signs posted outside.
(Figures 62 and 63) Long Beach and Compton had many unreinforced brick buildings before the M6.3 earthquake of 1933 and virtually all were severely damaged or demolished by the moderate shock. Most fatalities in American earthquakes so far are attributed to the collapse of such structures.
(Figures 64 and 65) Fifty years after the Long Beach earthquake, Coalinga's historic and mostly brick downtown was nearly destroyed in 1983.
(Figures 66 and 67) The M5.9 1987 Whittier earthquake in the Los Angeles area destroyed much of downtown Whittier (figure 66) and severely damaged brick buildings in Pasadena, Pico Rivera, Alhambra, Rosemead, Vernon, and many other towns. The Loma Prieta earthquake destroyed the beautiful Pacific Garden Mall and downtown Santa Cruz (figure 67).
(Figure 68) Damage from the M6.3 2008 Wells, Nevada, earthquake. The damage was primarily to older homes (chimneys) and URM commercial buildings.

(Figure 62)

(Figure 63)

(Figure 64)

(Figure 65)

(Figure 66)

(Figure 67)

(Figure 68)

CHAPTER 6

by URM buildings. In the 1994 Northridge quake, the overwhelming majority of URM buildings suffered significant damage, including complete collapse, collapsed walls, fallen parapets, and severe diagonal cracking. In the 2001 Nisqually, Washington, earthquake, two-thirds of the "red-tagged" buildings (those that were labeled with a sign indicating that they were not safe and could not be occupied) were of URM construction.

The 1933 Long Beach quake, which caused extensive damage to URMs, brought an end to the construction of such buildings in California, but not anywhere else in earthquake-prone areas of the United States. In 1986 California enacted the comprehensive "URM Law" to address the hazards posed by unreinforced masonry buildings. It applies to all jurisdictions in California's Seismic Hazard Zone 4, the areas of highest earthquake risk, which include the Los Angeles and San Francisco areas. The URM Law required local jurisdictions to inventory their URMs by 1990 and adopt a risk-mitigation program. Despite good intentions, most of California (and all other U.S. regions) have been slow to minimize adequately their URM risks.

In general, little can be offered in the way of encouraging (or, more specifically, inexpensive) advice to the owner of an unreinforced masonry building. It is worthwhile, in our opinion, to save and reinforce as many historically important URM buildings from demolition as possible. Old URMs are expensive to reinforce: the cost of such work often exceeds 25 to 50 percent of the value of the buildings. Still, many thousands of buildings have been strengthened. A walk through downtown Berkeley, San Francisco, or Los Angeles would take you by many of these now-much-safer buildings. Often, the most obvious evidence is the diagonal steel braces just behind the front windows.

One of the more common strengthening techniques entails removing the outer layer of bricks and replacing them with a layer of gunite (spray-on concrete) reinforced with steel bars. Gunite is certainly not as attractive as brick, but it can strengthen the walls adequately against collapse. At the same time, connections between brick walls and the floor and roof joists are strengthened with steel anchors that are driven through the brick and embedded in the gunite. Often it is also necessary to remove parapets and even the upper floors so that the repaired building can meet new earthquake code requirements.

Another common technique is the construction of a steel frame within or around the existing building. It is intended to provide an earthquake-resistant system that will keep the floors from collapsing even if the brick walls are severely damaged. It is very difficult to tie all of the brick to the steel, and much of it may come off, leaving a damaged and seriously disfigured building. More advanced techniques exist, too, such as the use of fiber-wrap reinforcement, which confines the damaged bricks.

URM buildings present a serious dilemma involving historical conservation and the preservation of older buildings to house the poor. For example, much of the charm of San Francisco's Chinatown is due to its old buildings—most of which are of unreinforced masonry, and many of which are densely populated by recent immigrants, elderly people, and others who cannot afford newer housing. However, a large percentage of these buildings will collapse or be seriously damaged in a future strong earthquake. Such a large-scale disaster is only a matter of time; it will occur unless these buildings are strengthened or eliminated. All other cities in earthquake-prone areas of the United States and British Columbia have the same problem in their historical centers.

Do not purchase or rent an unreinforced masonry building until you have consulted a structural engineer who specializes in building renovation and earthquake hazard abatement.

(Figure 69)

(Figure 70)

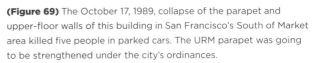

(**Figure 69**) The October 17, 1989, collapse of the parapet and upper-floor walls of this building in San Francisco's South of Market area killed five people in parked cars. The URM parapet was going to be strengthened under the city's ordinances.

(**Figures 70 and 71**) Friuli, in northern Italy, seen before and just after the 1976 earthquake. The destroyed buildings are mostly of unreinforced masonry.

(**Figure 72**) Unreinforced masonry buildings can be retrofitted to be safe, as is shown in this building that is located very close to an active fault.

(Figure 71)

(Figure 72)

DO UNREINFORCED MASONRY STRUCTURES STILL EXIST IN CALIFORNIA'S EARTHQUAKE COUNTRY?

Yes. Although significant progress has been made, about 25,000 URMs, in various stages of retrofit, remain throughout the state in areas designated as Seismic Zone 4 (the highest seismic zone). Most URMs house commercial uses. The following statistics (taken from a 2003 survey for the San Francisco Bay Area) indicate that a vast majority of URMs had not been made compliant with current codes. The problem is much worse, proportionally, outside California, especially in places like St. Louis, Missouri.

MAJOR CITY

SAN FRANCISCO
- URM Buildings   1,832
- Compliant with Current Code   1,018
- Demolished   117
- Percent Code-Compliant or Demolished   62 percent

OAKLAND
- URM Buildings   1,612
- Compliant with Current Code   1,329
- Demolished   108
- Percent Code-Compliant or Demolished   89 percent

BERKELEY
- URM Buildings   729
- Compliant with Current Code   612
- Demolished   6
- Percent Code-Compliant or Demolished   85 percent

SAN JOSE
- URM Buildings   146
- Compliant with Current Code   103
- Demolished   12
- Percent Code-Compliant or Demolished   79 percent

REGION (COUNTIES)

EAST BAY (Alameda, Contra Costa)
- URM Buildings   3,166
- Compliant with Current Code   2,112
- Demolished   155
- Percent Code-Compliant or Demolished   72 percent

NORTH BAY (Marin, Napa, Solano, Sonoma)
- URM Buildings   1,006
- Compliant with Current Code   218
- Demolished   23
- Percent Code-Compliant or Demolished   24 percent

PENINSULA (San Mateo, Santa Clara, Santa Cruz)
- URM Buildings   662
- Compliant with Current Code   388
- Demolished   62
- Percent Code-Compliant or Demolished   68 percent

(Figure 73) This figure shows an entire street of rubble construction destroyed by the Bhuj, India, earthquake of 2001. Brick and rubble construction is common in developing countries, which makes earthquakes in these countries usually more catastrophic with high death tolls.

Cavity-Wall Brick Buildings

As the name indicates, these buildings are constructed with a double wall of brick separated by a small gap. There are many such buildings in older West Coast cities, but the largest number of cavity-wall structures is found in Salt Lake City, St. Louis, and other Midwestern cities. Even when the double walls of such buildings are tied together with anchors (which is not usually the case), they have practically no resistance to the lateral forces of a quake and are readily damaged. Consult an engineer specializing in earthquake reinforcement if you wish to tackle the formidable and expensive task of strengthening such a structure.

(Figure 73)

Nonengineered Stone, Rubble, and Adobe Buildings

These types of structures, particularly those without any reinforcement or bracing, are able to carry only vertical loads and have practically no resistance to the lateral forces of earthquakes. The high casualty figures for earthquakes in South and Central America, southern Europe, and Asia are due primarily to this type of construction. During the Gujarat, India, earthquake of 2001, and the Wenchuan, China, earthquake of 2008, for example, entire villages were leveled, leaving thousands of people trapped beneath their destroyed homes and buildings.

RANKING YOUR BUILDING

Among small residential structures, the safest kinds of construction are steel and wood frame. For commercial structures and large residential structures the following applies:

Safest: Steel-framed buildings (such as modern high-rises). Stay inside these buildings during an earthquake—risk is greater outside the building due to falling windows and other façade elements.

Second best: Concrete shear-wall buildings; these rely on massive concrete walls rather than just concrete frames for structural integrity. Make sure there is no soft-story problem.

Usually OK: Modern (post-1976) reinforced concrete frame buildings. Older concrete can be very dangerous.

Sometimes OK: Concrete tilt-up buildings (primarily if built since 1976 or retrofitted), which are seen in many industrial parks. Weak tilt-ups are typically relatively inexpensive to strengthen.

Dangerous: Nonductile concrete-frame buildings, which are often older structures. Especially prevalent in San Francisco are six-story early-twentieth-century multistory apartment buildings.

Think about moving or changing jobs: Unreinforced masonry buildings have caused most deaths in recent U.S. earthquakes. If you live or work in one, make a plan to leave. The same applies to many older concrete buildings.

Steel framing for a two-story home

CHAPTER 7

Structural and Architectural Design for Earthquake Resistance

The most earthquake-resistant building would be a clumsy, boxlike structure with a minimum of windows and doors penetrating its walls. Numerous interior partitions dividing the house into small rooms would further strengthen the structure. Obviously, such a building would not be very appealing, nor is it very common. Yet it is possible to achieve architectural beauty and earthquake resistance in the same structure. As we will see in this chapter, earthquake resistance does not need to mean a bunker-style home or building.

Until recently, when the need for energy efficiency began modifying the architectural appearance of houses, contemporary design trends favored buildings without any visible means of support, and certainly such light and airy structures are attractive and well suited to the climate and lifestyle of California and other warm-weather locations. Such buildings often have large glass areas, slender columns supporting broad-eaved roofs, multilevel floor plans, unusual geometric patterns, and dramatic water and hillside sites.

Unfortunately, the design that makes these structures attractive also incorporates many of the most obvious earthquake hazards. Engineers sometimes neglect to stress the importance of good detailing, and contractors do not have the knowledge to insist on proper earthquake-resistant design. Consequently, a large percentage of building damage from earthquakes is directly attributable to poor design or detailing.

For example, the failure of a slender corner column supporting a heavy roof or an upper floor in a split-level home can damage not only walls and large glass panels but also the roof, floor, interior walls, furnishings, and occupants of the building. In contrast, a sturdy, simple home with conventional windows and numerous supporting walls might sustain some sheetrock cracks and chimney damage, but chances are very good that it would remain intact, and the damage would not be costly or dangerous to occupants.

As this book has continually emphasized, the modern design requirements for houses and other buildings in earthquake country must be accompanied by special attention to the strengths of the load-bearing and earthquake resistant elements and the connections among all the structural elements of the building. This chapter examines these factors in greater detail and suggests ways that poor detailing and design can be avoided in new construction or corrected in existing buildings. It also discusses the special earthquake problems of older buildings and of certain modern architectural features that caused or sustained disproportionate damage during recent earthquakes.

Foundations

A good, reliable foundation is a continuous tied wall foundation, in which reinforced concrete provides a uniform length of support under the main load-bearing components of the building. The concrete is reinforced with horizontal steel bars that wrap around corners and is securely tied together with vertical steel bars. Such tied foundations enable a building to move as a single integrated unit during an earthquake, so that different components of the superstructure move together and damage is minimized.

A mat, or floating, foundation—a reinforced-concrete slab resting directly on the soil—is ideal for buildings on soft soils or other inferior ground, such as landfill. Full mat foundations, when well reinforced with steel, have the advantages of rigidity and continuity—they provide continuous support to a structure and minimize the hazard from differential soil movements by bridging over pockets of especially soft or loose soil. Unfortunately, when foundations displace even slightly, the support of the building becomes uneven, causing severe cracking and warping of the floor. This can lead to damage in the frame. Buildings on mat foundations (and those on the pier or piling foundations described below) generally experience less of this type of damage during earthquakes. Much of the damage in the Marina district of San Francisco during the 1989 earthquake would not have occurred had thick mat foundations been used over the soft fills of the area.

Drilled pier or caisson-pile foundations, which are steel or concrete pilings set deep into the ground, are generally used only on soft, weak, or unstable soils. These foundations have a good record for endurance in large earthquakes. For example, much of the damage in the alluvial and filled areas of San Francisco in 1906 was caused by large ground settlements. However, buildings on deeply submerged pilings did not settle with the alluvial or filled soils, and damage to these buildings was substantially less. San Francisco's Ferry Building is one of those survivors.

The quality and earthquake resistance of these three types of foundation are governed by a few basic principles:

- The foundation should be supported by solid ground, and its major supporting segments should rest on uniform ground conditions. For example, a foundation wall should not be partially supported on bedrock and partially on landfill material; differential movements of the fill away from the rock during a quake can damage the foundation and the structure it supports.
- Different types of foundation should not be used under one building—a combination of separate unconnected pilings with sections of continuous concrete-block wall, for example, could be risky unless the design engineer thoughtfully anticipates the earthquake hazard. Two different types of foundation will move in different ways during the ground motion of a tremor, and uniform behavior is essential.
- The type of foundation required for a given site should be determined primarily by ground conditions under the building. A good bedrock base, or well-compacted alluvium, needs only the conventional continuous tied-concrete foundation minimally required by most building codes, whereas very soft soil or loose fill may require special drilled pier or caisson footings.

(Figure 1)

(Figure 1) This is an example of a well-tied and continuous wall foundation for a two-story house. Note the sill anchors (bolts) every few feet and the taller shear wall tie-downs which will connect the foundation directly to the plywood sheathed walls above.

(Figures 2 and 3) Details of an unreinforced brick foundation wall that has been strengthened with reinforced concrete after some bricks have been removed.

GETTING GEOLOGIC AND ENGINEERING HELP

There is a tremendous amount of misinformation about what constitutes professional help in evaluating and strengthening your house against earthquakes.

Two professional groups are equipped to help: licensed engineers and geologists. These professionals are licensed by their state. Most others are either unqualified or inadequately qualified to dispense such critical advice. Only engineers, geologists, and some architects are trained in the fundamentals of earthquake engineering for small buildings. Carpenters, builders, and contractors are not usually trained in this field.

When you need help, seek either an engineer (civil, structural, geotechnical) or a geologist. Their advice can aid you in a variety of situations—purchasing property, strengthening through structural alterations and additions to your house or building, or dealing with problems of unstable ground, landslides, settlement, and the like.

We strongly urge everyone buying a house to seek both engineering and geotechnical advice about earthquake safety before making the purchase. Generally, you'll need two reports, because the expertise of structural engineers and geotechnical engineers does not overlap. It is best to include a brief clause in the contract that makes purchase of the house contingent on these reports; as the buyer, the one who inserted the clause, you can remove it from the contract whenever you want. Allow yourself five to ten working days to find an engineer. You might consider contacting one when you start looking for a house.

If you're seeking an educated opinion rather than an escape clause, you will not usually need these reports in writing—which is something that can increase the cost. Just ask for oral advice and simple sketches that show you how to accomplish the specific recommended strengthening details.

In our opinion, such inspections are excellent investments. They can save you serious problems in the event of an earthquake—or prevent you from buying a house on top of a future landslide.

To find an engineer, ask any architect or engineer you know, call your city's engineering office, or check with your real estate agent. We advise against simply doing a general Internet search for help. Another good source for engineers is your state's Structural Engineers Association (SEA), which maintains lists of engineers who conduct such evaluations. You can access all SEAs by going to www.seaint.org. The email address is seaint-ad@seaint.org. In all cases, make sure that the person you hire is a licensed geotechnical, civil, or structural engineer in the state of your residence.

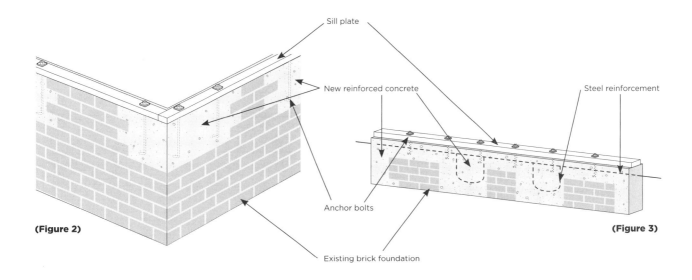

(Figure 2)

(Figure 3)

Foundation Connections for Wood-Frame Buildings

One of the most common causes of costly damage to wood-frame buildings—which is also one of the easiest problems to fix—is insufficient or poor anchorages between the sill and foundation.

The Carnegie Institution report on the 1906 earthquake notes that many damaged single-family houses in San Francisco had "slipped from their foundations." Houses in almost all the cities on the northern coast of California suffered similar damage. This was also the most common severe home damage observed following the 1989 San Francisco and 1994 Northridge earthquakes and all other western-state earthquakes since 1906. Building codes now require firm anchorages between the sill and foundation in all regions of the United States—whether seismically active or not. For earthquake-prone regions, more stringent criteria are enforced.

Any enterprising homeowner can bolt down a house. It may take a day or two, or it may take three weekends. Or you can hire help.

First, check whether the wood sill underneath the walls is bolted to the concrete foundation below. The sill is a thick (2- or 3-inch) wooden plank on top of the foundation to which the vertical studs of the house walls are nailed.

To check, go into the basement or crawl under the house and inspect the foundation. You will see the steel washers and nuts of the bolts if the sill is bolted to the foundation. Most pre-1940 houses do not have such bolts. All post-1940 houses should have them, but you should check, just in case the builder decided to cut costs or was uninformed.

If there are no sill bolts, install them according to the sidebar procedure on page 118. Use a wood bit to drill through the wood sill; then use a rotohammer with a carbide-tip bit for the concrete. You can rent this kind of drill. If the space is too low for the drill, use a right-angle drill attachment. You can find foundation bolts and washers at any major hardware supplier.

The sill bolt used for concrete is called a wedge bolt (or sometimes an expansion bolt). The hole drilled into the sill and foundation does not have to be precise in its length, but it should be longer than the bolt. The diameter of the hole should provide a snug fit for the bolt. The hole should be cleaned by blowing out the debris inside with a small-diameter hose or a straw.

Wedge bolts can crack older and more crumbly concrete and masonry foundations. In these cases, chemical anchors (also called epoxy bolts) are the preferred option. In this case, be careful not to drill much deeper than the bolt's length. Then inject the epoxy mixture into the hole. Press the bolt into place and wait for the epoxy to harden (usually twenty-four hours). Once the epoxy has hardened, tighten the nut with an adjustable wrench until the washer just begins to indent the wood sill. Chemical anchors can be difficult and time consuming to install but also can be very effective. Other types of proprietary anchoring systems are also available, but designing and installing them requires the help of a structural engineer.

If there is no access to the sill from the crawl space, the sill can be bolted from the exterior by removing the exterior coverings, such as wood siding, and bolting flat steel plates into the sill and concrete foundations.

If the foundation is made of brick rather than concrete, consult a civil or structural engineer specializing in earthquake-hazard reduction. Usually, the engineer will recommend removing at least part of the brick foundation and replacing it with a concrete one (see page 113). This can be expensive.

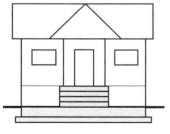
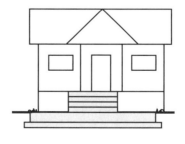

(Figure 4)

Crawl-Space Wall Bracing for Woodframe Buildings

Weakly braced or unbraced underpinnings are the second most common weakness of older (pre-1950) houses, and of many newer houses that are supported on crawl-space walls. Typically, these framed 2-by-4-foot crawl-space walls extend 2 feet or less between the concrete foundation and the flooring, but in some houses they are taller. In houses on sloping ground, they may be 8 feet or higher.

A very high proportion of earthquake damage to wood-frame houses in the United States is caused by weak crawl-space walls. The weakest ones are those that are covered on the exterior of the house with lateral wood siding or shingles. Do not be fooled by these exterior finishes: they provide no effective lateral (or shear) strength. Instead they allow crawl-space walls to lean to one side and collapse. Exterior stucco applied directly over the studs is a little stronger than straight siding, but it is not strong enough to prevent damage.

Houses with weak crawl spaces have plenty of walls above to absorb the force of shaking. But in their crawl spaces, only the peripheral walls have any strength; there are generally no other interior walls to help—only short columns on individual footings and a few framed walls without any sheathing.

In both the 1989 San Francisco and the 1994 Northridge quakes, several hundred crawl-space collapses occurred in the area of the epicenter—and most were easily avoidable. In some neighborhoods, entire blocks of older houses collapsed. Some of their foundation sills were bolted and some were not, but it did not matter, because their crawl-space walls collapsed.

See sidebar on page 120 for how to add bracing. Before adding bracing, make sure that the house is bolted to its foundation.

At least 50 percent of the length of each cripple wall should be braced for a single-story home; 80 percent should be braced for two-story homes. Avoid using multiple short pieces of plywood, and try to have as few breaks in the plywood as possible.

Usually, it is further recommended that you add at least 8 linear feet of plywood bracing in each interior corner of the crawl space in each direction. Thus, if the crawl space is 3 feet high, you need to strengthen the corners of your house with eight plywood sheets, each 3 feet high by 8 feet long.

If the crawl space is higher than 4 feet, then the plywood length on each face, in each corner, should be at least twice as long as it is high. Thus, if the crawl space is 5 feet high, you should use a total of eight sheets, each 5 feet high by at least 10 feet long.

Costs and Benefits

The cost of the materials for bolting the foundation and bracing the crawl-space walls is usually less than $1,000. If you hire a contractor, expect to pay upward of $5,000 for a typical house and upward of $10,000 for a large, difficult-to-fix house.

Were all owners of older wood-frame houses to bolt their sills and brace their crawl-space walls, we estimate that 75 percent of all serious earthquake damage to such houses would be eliminated. Loss of life, too, would be dramatically reduced. We strongly urge you to take these two steps. Unfortunately, only 10 percent of homeowners have taken these steps.

(Figure 4) Illustration showing a house that has not been correctly attached to its foundation being seperated from its foundation due to earthquake shaking.

(Figure 5) Illustration of a crawl-space wall failing due to earthquake shaking.

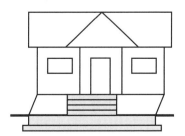
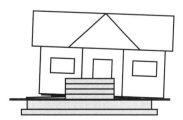

(Figure 5)

(Figure 6)

(Figure 7)

(Figure 8)

(Figure 9)

(Figure 10)

(Figure 11)

(Figure 12)

(Figure 13)

(Figure 14)

(Figure 15)

(Figure 6) Contemporary code requirements provide for connections between the foundation sill and the walls of a building. However, older buildings—some constructed before the mid-1950s and most built before the 1940s—often have no connections between the foundation and frame. This house slid off its undamaged concrete foundation in the 1971 San Fernando earthquake.

(Figure 7) The foundations of older houses are unlikely to have any significant connections or bracing between the foundation and the structure. Witness this dramatically ruined Victorian house in San Francisco in 1906. The force of the fall from its foundation split the building in half.

(Figures 8 and 9) On several streets in Watsonville, every house failed during the 1989 Loma Prieta earthquake (figure 8). For all of these, either they were not bolted down to their foundations and/or their cripple studs were not braced with plywood. A few houses survived—this one (figure 9) had been bolted and otherwise strengthened shortly before the earthquake. It had no damage.

(Figures 10 and 11) The crawl-space wall of a house in Watsonville failed because the walls were not braced (figure 10). The foundation of a neighboring house (figure 11) had been removed and a new concrete peripheral wall constructed. This house had no damage.

(Figures 12 and 13) The plywood sheathing of the improperly nailed cripple wall of this house failed during the M6.0 Morgan Hill, California, earthquake of 1984. Figure 13 shows the house during repairs, after it was leveled and new plywood was applied.

(Figures 14 and 15) Two houses in Santa Cruz after the October 17, 1989, earthquake. The house shown at in figure 14 had recently been renovated with simultaneous structural upgrades. The house in figure 15 had not been strengthened and came off its foundation.

BOLTING YOUR HOUSE TO ITS FOUNDATIONS

1. Lay out bolt locations.
2. Drill holes through existing sill into the concrete foundations using carbide drill bits and an impact-type drill. Use right-angle drill for tight access places where the crawl space is low.
3. Blow all the dust out of the drilled holes using a rubber tube.
4. Insert expansion* bolt with the washer and nut attached. Leave nut at top of bolt when tapping the bolt in place to protect the threads.
5. After tapping bolt in place, tighten the bolt by turning the nut. Do not over tighten or bolt will be damaged.

* epoxy bolts may be more appropriate in older foundations. Follow manufactures installation instructions.

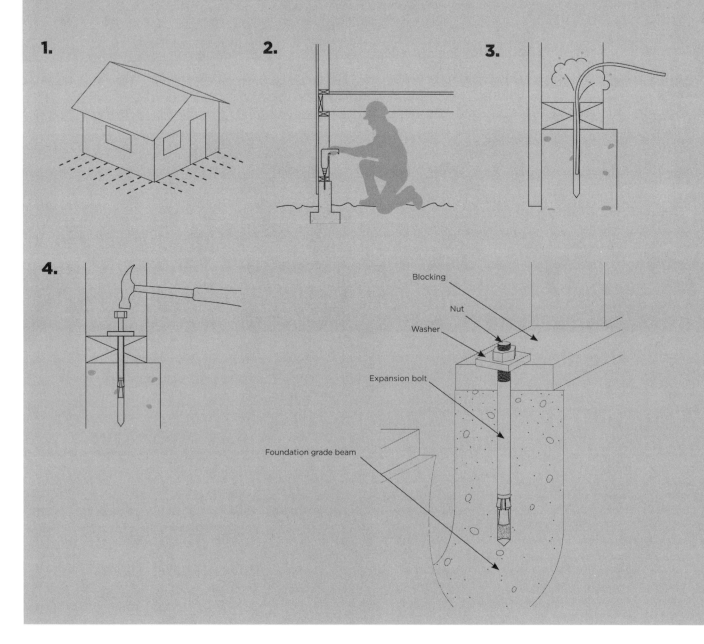

- Sill plate should be bolted to the foundation with 5/8-inch-diameter steel bolts (or 1/2-inch-diameter bolts in areas of low seismicity).
- Bolts should be embedded at least 7 inches into the concrete or masonry.
- For two-story structures, bolts should be spaced not more than 4 feet apart.
- There should be a minimum of two bolts per sill piece, with one bolt located not more than 12 inches or less than seven bolt diameters from each end of the piece.
- Plate washers with minimum dimensions of 3 inches by 3 inches by 1/4 inches thick should be used on each bolt.
- In a structure taller than two stories, the sill-to-foundation connection should be individually designed by an engineer.

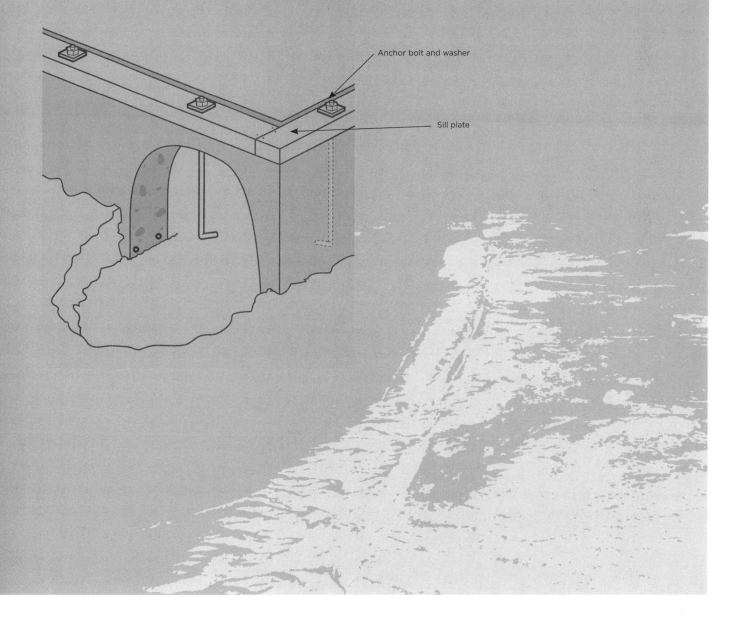

Anchor bolt and washer

Sill plate

SHEATHING YOUR CRAWL-SPACE WALLS

1. Check the sill plate to be sure it is adequately bolted to the foundation (see sidebar on page 118).
2. Check to be sure the sill plate and studs are the same dimensions (flush at face). If not, block between the studs, and nail the blocks into the sill plate with four to six 10d nails per block to create a flush nailing surface for the plywood. Block around the foundation nut and washer.
3. Measure the distance between the bottom of the sill plate and the top of the top plate. Measure the distance between the corner stud and the middle of a stud at 4 by 8 feet away to be sure a standard sheet will fit; if not, you will need to cut two sides of the 4-by-8 sheet. Check that the studs are square with a framing square, and use it to lay out cuts on the plywood.
4. Mark the center line of the vertical studs on the foundation and above the cripple wall to locate studs behind the plywood for later nailing.
5. Locate all exterior ventilation grates in relation to an easy reference point so that you do not cover them with plywood.
6. Cut the plywood, using a circular saw with a plywood blade.
7. Place each precut section of plywood up to check its fit. It may need to be trimmed; you can use a jigsaw to trim it without taking it out of the crawl space.
8. Temporarily tack up the plywood with a few nails. Using a chalk line, snap a line on the plywood between the marks made in step 4. Nail the plywood to the studs and plates with 8d nails. The nails should be spaced 3 inches apart around the entire perimeter of each plywood panel and 6 inches apart in the middle of each sheet. Stagger the nails at plywood joists.
9. Using the jigsaw, measure and cut out a space of the same dimensions at the ventilation grates that you located in step 5.
10. Drill two 2½-inch to 3-inch ventilation holes for each cavity between the studs. The holes should be 2 inches up from the sill plate, 2 inches down from the top plate, and centered between the studs.
11. Measure the next section to be cut and fit only after you have completely attached the previous section.

Below, shows the proper placement of the plywood sheathing. Panels should be placed at both ends of each crawl-pace wall section. At least 50 percent of the length of each crawl-space wall should be braced in a single-story home; 80 percent should be braced in a two-story home. Homes beyond two stories should be engineered by a structural engineer. Use the longest piece of plywood possible; avoid using multiple pieces of plywood. Use ½-inch (or thicker) five-ply CDX structural-grade plywood. (Do not use "shop-grade" plywood.)

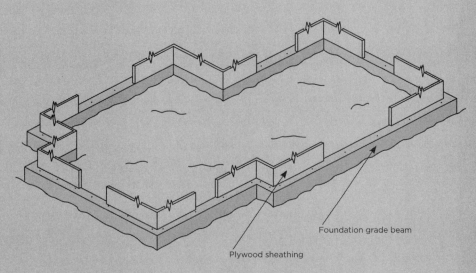

Foundation grade beam
Plywood sheathing

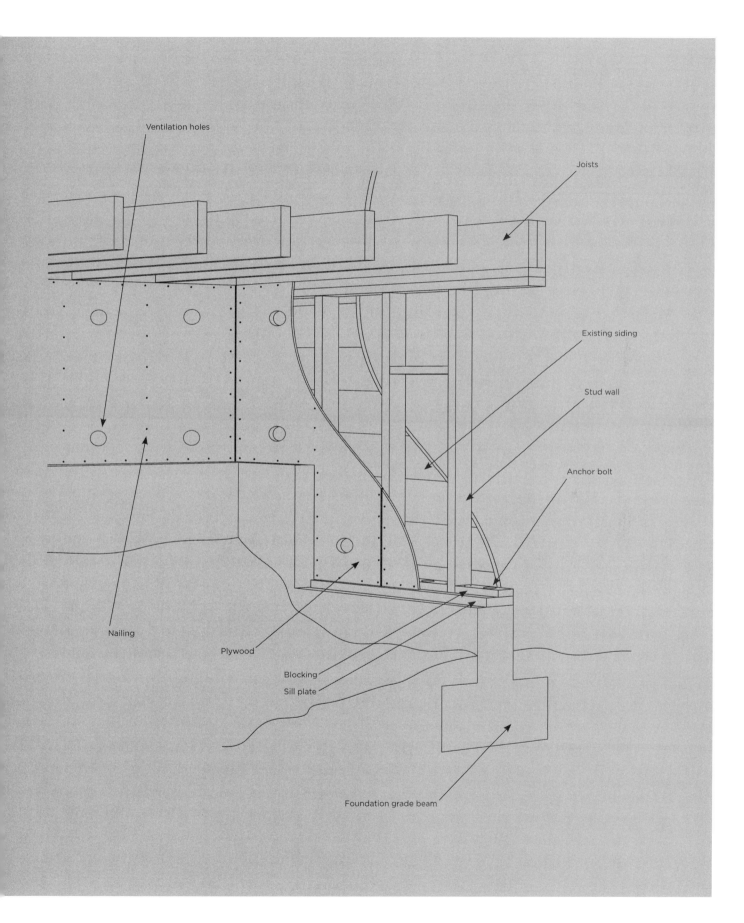

Existing Foundation Damage

Older buildings, especially, are likely to have some foundation damage and should be carefully checked. When inspecting for foundation damage, give special attention to the condition of the wood connections. The foundation may have been laid or graded improperly so that it traps water and allows rotting in the wooden sill or columns. Poor insulation or an improperly placed dirt backfill may also retain or trap moisture and rot the wood connections. Termite damage to the sill or columns is another particularly hazardous factor. The buyer of a home should always insist on a termite clearance, even when it is not required.

Wide cracks (more than 1/8 of an inch) in foundation concrete are an almost certain sign of existing damage from ground settlement. (Do not confuse such cracks with common—and harmless—"hairline" cracks, which are caused by the shrinkage of concrete.) These wide cracks may also indicate further structural damage in the walls, so it is imperative that you consult an engineer. Existing foundation damage must be repaired in most cases, since not only is it an earthquake hazard but it may also lead to serious damage to the structure simply in the course of time. The repairs can be expensive, so be sure to shop around among contractors to get the most competent and economical work.

Houses on Stilts or Pilings

The construction of houses on stilts or pilings has made it possible for people to enjoy living on steep hills or along bay shores and riverbanks. Such houses are very appealing, but they also can be very dangerous. As noted, the ideal foundation support for a house in earthquake country is a continuous tied wall of reinforced concrete or a reinforced concrete slab. Houses founded on a few slender supports are too seldom designed to carry the large lateral forces of an earthquake. Earthquake risks for this type of building are compounded by the propensity of shoreline and steep hillside soils to amplify earthquake intensities or slide.

A house can be safely supported on slender columns only when its soil foundation is exceptionally stable and its columns and their piers below ground are specifically designed and braced for earthquake loads. For example, columns and floor joists of welded steel are one excellent solution for a steep hillside house. The columns must be embedded in deep concrete piers and the floor joists must be thoroughly bolted and welded to the columns.

For the more-conventional and less-expensive wood columns, the safe choice is a shear-wall bracing system of plywood covering the entire length and breadth of the foundation. Diagonal bracing is never adequate for timber stilts, and if you own such a house you should certainly add continuous plywood sheathing as soon as possible. Consult an engineer to ensure that the design and mounting of the bracing complies with sound engineering principles.

Buildings partially or wholly on raised pilings are also very susceptible to severe damage. If a house is supported entirely by pilings, they must be long enough and securely planted in the ground and rigidly tied to the understructure of the building. The ties will prevent movement and the consequent spreading and collapse of the pilings.

(Figure 16)

(Figure 17)

(Figure 18)

(Figure 16) Houses on stilts on very steep hillside sites can be a poor investment unless very carefully designed by a structural engineer.
(Figure 17) Shown here is one such house collapsed by the Northridge earthquake.
(Figure 18) Shear-wall bracing of plywood, applied over conventional studs, should be used for a steep site, as in this house shown here.

Columns, Walls, and Split-Level Houses

In general, the failure of a wall or of the main corner columns of a building is one of the most serious types of earthquake damage. Once an important wall fails or a column breaks or becomes disconnected, a chaotic sequence of further failures begins that can demolish the structure.

Wood columns fail most often because of the following deficiencies:

- The column is weakened by rotting because of poor drainage along the top of the foundation. Termite damage to the column is another common factor. If any rotted or termite-infested column bases are found in a home, immediately replace the column, making careful provision for better drainage and thorough treatment of the wood against water or insect damage.
- The joists, sills, and other horizontal structural members are inadequately tied to the column. This is by far the most common and troublesome weakness of columns in enduring earthquake forces. It is also the easiest and cheapest to correct. The liberal use of steel angles, either when building a new home or when reinforcing an existing structure, adds tremendous strength to the typical wood-frame building.

Split-level and multistory houses or apartment buildings with garages at ground level are widespread in earthquake country. Most such buildings were constructed after the early 1950s, and, in the stronger California earthquakes, they suffered a disproportionate amount of the total damage, as did houses with lots of glass and commercial buildings, such as stores, with all-glass fronts on the ground floor. These buildings are often called soft-story buildings. The apartment buildings that collapsed in the Marina district of San Francisco in 1989 and in Los Angeles in 1994 were all of this type.

These buildings are inherently weaker than conventional buildings because openings on the lower level weaken a portion of the wall area that must carry and resist earthquake forces. The garage level becomes, in effect, a foundation with only three walls. During an earthquake, such a building tends to twist. But an essentially three-sided structure cannot easily withstand this twisting, particularly under the additional weight of another floor or two above it. If the garage walls do not eventually give way entirely, their top portion and the corners of the garage doors and windows may be severely cracked. In this case, the whole building is also susceptible to further damage or collapse should a strong aftershock occur.

This type of damage has occurred most often to houses without adequate shear-wall bracing. To minimize such damage and prevent collapses, garage walls, especially front and back walls, should be strengthened with plywood sheathing. Wallboard sheathing and diagonal bracing are totally inadequate for strong earthquakes. It is also very important to use steel hold-downs or other anchoring devices to securely connect the plywood to the foundation sill and the floor joists above. Any columns that support intermediate floor joists should also be thoroughly connected with steel connectors.

(Figures 19 and 20) Shown here are two of the numerous split-level and two-story homes in San Fernando in which lower-level garage walls were inadequately braced for the heavy load they had to support. The right ground-level side of the first house also had diagonally braced cripple studs, which contributed to the devastation. In the second photograph, note that the upper level of the house simply ripped away as it crushed the garage walls and the car inside. No portion of the house seems to have suffered any other structural damage. Plywood sheathing on the garage walls of a split-level can mean the difference between minor damage and costly destruction.

(Figures 21 and 22) Before-and-after photographs of a soft-story concrete apartment building in Santiago, Chile, in the M7.8 1985 earthquake. A few concrete shear walls, properly placed on the ground floor, would have prevented the collapse.

(Figures 23 and 24) This soft-story design for modern apartment buildings, very common on the West Coast, is both risky to the investor and unsafe for the occupants unless proper measures are taken to strengthen the supporting walls and the connections between the walls of the different stories. With their essentially three-sided supporting foundation and thin columns, these San Fernando buildings were fortunate indeed to still be standing after the quake. A stronger shock or a few more years of inevitable deterioration probably would have meant their collapse. All of the collapsed buildings in San Francisco's Marina district during the 1989 earthquake were of this type.

(Figure 19)

(Figure 20)

(Figure 21) Before

(Figure 22) After

(Figure 23)

(Figure 24)

(Figure 25)

(Figure 26)

(Figure 27)

(Figure 28)

(Figure 29)

(Figure 30)

(Figure 31)

(Figures 25–27) Most damage to houses in the Marina district of San Francisco during the 1989 earthquake was due to weak lower stories (figure 25). Typically, the first stories of these houses were completely open, without any structural cross walls (figure 26) or plywood bracing at the garage door or in the back. Plywood bracing, such as that being applied to the houses in figure 27, would have prevented the damage.

(Figures 28–30) Three different ways of strengthening weak two-story houses in San Francisco's Marina district. Figure 28 shows plywood sheathing applied on all walls parallel to the street (the weak direction of the house). In figure 29, plywood will be applied to the diagonally sheathed wall. A new backup wall has already been built behind (with plywood sheathing on both sides). Figure 30 shows a steel frame added in the middle of the garage. The steel frame is more expensive but does not take up significant space.

(Figure 31) Steel framing is again shown for this new home under construction.

WHAT IS A SOFT-STORY?

A soft-story is created when the first story (usually) is left open for parking, a shopfront, or other purposes. The stiffness of this story is significantly less than that of the stories above and will therefore suffer more movement and damage in an earthquake, quite often resulting in collapse. Collapse of the first floor can lead to progressive collapse (pancaking) of the entire building.

As the following schematic illustrates, the building without a soft-story on the left is able to distribute the building displacements throughout the structure. For the building with a soft-story (center and right), the first story takes the majority of the displacement and subsequently fails. The two images below further illustrate this point. The building on the bottom left, unless retrofitted, would be considered a soft-story structure. A similar structure, shown at bottom right, collapsed after the 1994 Northridge earthquake. (The lower floor is missing).

The top row of images on the facing page shows damage during the 1994 Northridge earthquake.

The middle row of images illustrates the important point that garage doors do not provide earthquake resistance, although they provide the illusion of a continuously braced building. Although visually, it looks as if there is a wall, the garage-door will not protect the building. To check if the building is retrofitted, look behind the garage door as shown in the photograph on the middle right. If the building is retrofitted, you would see a steel frame or other means of bracing the lower story. (This was not the case in the garage in the photo). Note also the potential soft-stories just across the street.

The bottom row of photos shows good examples of buildings that are not soft-stories. Note the continuous structure through to the ground.

If you think you live in a soft-story building, seek the advice of a structural engineer or request that your landlord provide you with an engineering assessment report.

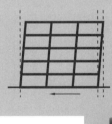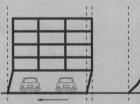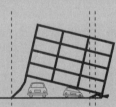

CHAPTER 7

Follow these guidelines:
- If the garage extends the entire depth of the house, so that there is no complete wall parallel to the garage door between the front and the back of the house, build a new wall (with its own reinforced concrete footing) to block off part of the garage.
- If a wall parallel to the garage door exists, separating the garage from another room in the back of the house, such as a family room on the lower floor, sheathe that wall with plywood.

If you own a stuccoed wood-frame building without plywood sheathing, take one or more of the following steps:

- Add interior or exterior plywood shear walls at selected locations to stiffen and strengthen the house. This measure should not only reduce fractures around the windows and doors but also minimize damage to the rest of the building's stucco covering, since window and door fractures frequently extend into a web of cracks and breaks all across the building face.
- If your building is taller than one story, add plywood sheathing at the ground floor (or, better, on all floors).

Sometimes a steel frame can be inserted behind or around the garage door. This is by far the most effective way to avoid the pitfalls of soft-story construction in new buildings. In fact, adding a steel frame around a garage opening in two-story houses has become quite commonplace throughout California. For example, the authors recently observed new houses under construction in Manhattan Beach (in the Los Angeles area)—where lots are typically narrow and new houses are generally two or three stories with garages on or below street level—to include such frames.

(Figure 32) Buildings with heavy roofs, including clay-tile and slate roofing, require significantly stronger wall bracing. This building, at the Fukui, Japan, College of Engineering, vividly illustrates the whip effect caused by the combination of the heavy tile roof and the 1948 earthquake.
(Figure 33) This single-story dwelling in Managua, Nicaragua, endured the heavy load of its tile roof and a strong tremor with only light damage in the 1972 earthquake. Its strong walls and low level saved it from further damage. However, another disadvantage of tile roofing is shown here. Without careful anchoring, the tiles are easily thrown off the roof during an earthquake.
(Figure 34) Heavy tile roofs tend to cause a lot more damage than lighter roofs, as can be seen in this photograph from the M7.2 Kobe earthquake in 1995. All but the house on the right in the photograph have collapsed. The collapsed houses had clay tile roofs; the surviving house had a light composition-shingle roof.

Often the steel frame is extended above ground level to allow for large windows in the stories above. We highly recommend this detail in both new construction and retrofit strengthening. The steel frame can be further enhanced by adding manufactured shear panels made of steel. Consult a structural engineer.

Floor and Roof Diaphragms

Most earthquake failures associated with a structure's diaphragm members occurred at their connections with the walls, rather than in the diaphragms themselves. The destructive forces of an earthquake are easily absorbed by these horizontal members, but their connections to the vertical supports are less durable. This is a problem mainly in masonry and tilt-up concrete buildings with wooden or steel-frame roofs and floors. In an earthquake, the rigid masonry walls pull away from their connection with the diaphragms, and the suddenly unsupported roof or floors plummet.

(Figure 32)

(Figure 33)

Diaphragm connections can also fail because of the exclusive use of a wood ledger—a framing beam that is bolted to the masonry walls. The floors and roof rest on and are nailed to this ledger. Such a connection is insufficient in earthquake country; roofs and floors usually require the additional support of steel anchors or metal straps that tie the joists of the diaphragms directly to the wall. These anchors and straps keep the diaphragms from pulling out of the ledger and from pulling the ledger out of the masonry.

The most common roof diaphragm problem in wood-frame buildings is a heavy roof. Heavy tile roofs are ever more popular in Southern California, partially because they reduce fire susceptibility, and partially because of the proliferation of Mediterranean architectural styles. They are also popular in Japan and have contributed to many collapsed homes during recent earthquakes. Mission-style clay-tile roofs, slate roofing, and other heavy roofing materials require considerably more support and reinforcement to secure their connections during an earthquake than conventional roofs do. A clay-tile roof on a 1,500-square-foot building, for example, weighs about 8 tons more than a roof of wood or asphalt shingles. The heavy-roof problem is especially hazardous in multistory buildings because such buildings amplify ground motion in their upper floors, causing much larger inertial forces on the walls of the building. It is essential that the walls of a building topped by an exceptionally heavy roof are strengthened with shear-wall bracing of plywood. Even with such bracing, there is some risk of damage to or by the roof, and it may be a good idea to carry earthquake insurance.

Clay-tile, slate, and other masonry or stone roofing materials are also in themselves especially susceptible to earthquake damage. Unless they are nailed to the roof, they tend to dislodge and fall during even moderate tremors that leave the rest of the building completely unaffected. Besides being expensive to replace, these heavy roofing materials can be dangerous to people running out of the building during a quake.

(Figure 34)

(Figure 35)

(Figure 36)

(Figure 35) Similar heavy tile roof damage from the 2007 Niigata, Japan, earthquake.
(Figure 36) Roof strengthening. Note the original roof on the left. Well-nailed plywood strengthens the roof and provides additional continuity. This continuity provides a much stronger and effective path for the earthquake loads to travel (be transferred) to the various walls in the house under it. Shown here is Peter's roof being strengthened when the old tar and gravel needed to be replaced anyway.

THE PROBLEMS WITH IRREGULARLY SHAPED BUILDINGS

The plan, shape, and configuration of a structure will affect the way in which it responds to earthquake shaking. Although buildings with complex geometries can be designed to be safe, with all other parameters being equal, a structure that is symmetric and regular in shape will likely suffer less damage than one that is complex in shape. The reason for this is that vertical and horizontal eccentricities like those shown below cause forces to concentrate into certain locations, causing concentrated areas of damage. For regular buildings, the forces are more evenly distributed, generally resulting in less damage.

Vertical Eccentricities

Horizontal Eccentricities

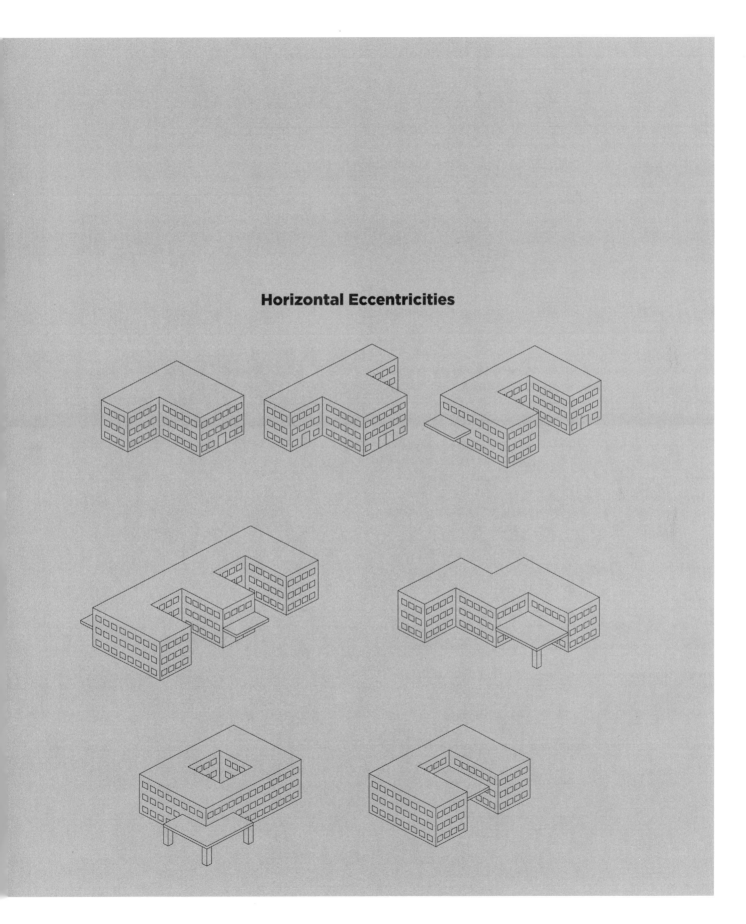

CHAPTER 7

Large Windows and Doors

These introduce weak points into the structural frames of buildings. Much quake damage is concentrated in their vicinity. Why is this? Where there are no openings in the walls, earthquake forces are distributed evenly throughout the entire wall. A large hole in the structural frame interrupts the path of the earthquake forces and increases stress on the area around the opening. The sharp corners of the opening also cause further stress concentrations that add to damage. Large plate-glass panels, such as sliding glass doors, bay and picture windows, and glass walls, present similar hazards.

A careful plywood bracing system around openings is sufficient to redirect earthquake forces back to the foundation and to stiffen and strengthen the framing. Shear-wall bracing should be placed below, above, and especially on either side of large windows or doors, and the sheathing on either side of the opening should be at least as wide as the opening itself to prevent cracking and other damage to interior and exterior wall veneers.

(Figure 37) Large windows and inadequate bracing caused this structure to collapse during the 1989 Loma Prieta earthquake.

(Figures 38 and 39) Large window and door openings are especially prone to earthquake damage because the openings interrupt the path of earthquake forces and concentrate stress around the openings, especially at their corners. Very often, the cracks that appear at these openings indicate far more serious structural damage within the walls. This is shown in this damaged structure in figure 39, from the 1984 Morgan Hill earthquake. Figure 38 shows the proper addition of plywood sheathing around all large openings on the ground floor of a house under construction.

(Figure 40) Masonry parapets and other overhanging, unsupported architectural features are highly subject to earthquake failure and are one of the most hazardous elements of earthquake damage. The parapet and the top portion of this unreinforced masonry building in downtown Whittier collapsed in the M5.9 1987 earthquake.

(Figure 37)

(Figure 38)

(Figure 39)

Parapets, Ornaments, Balconies, and Other Projections

Masonry parapets and other ornaments are usually the first components of a building to fail, and fall out, during an earthquake. Their positions at the tops of buildings, where earthquake vibrations are most intense, poor connections between them and the building, and the cracking and general weakening caused by weathering and lack of maintenance all contribute to the frequent failures of these architectural features.

In themselves, these decorative projections are not a large problem, because, if they fall or break, an owner can simply eliminate them from the building and repair costs will be moderate. However, when they fall, they are also likely either to pull down a part of the supporting wall or to damage the lower portion of the building or other adjacent buildings. In addition, because these architectural features tend to be located above entrances, they are extremely hazardous to people running out of buildings during a tremor. A large percentage of all deaths from earthquakes in California since the 1906 San Francisco tremor were caused by the collapse of brick parapets. The only deaths in smaller recent California earthquakes, like the magnitude 6.5 San Simeon earthquake in 2004, were caused by falling brick parapets.

Fortunately, the repair and strengthening of hazardous projections are usually moderately inexpensive. Many but not all cities in California have passed retroactive building ordinances requiring the elimination or strengthening of hazardous parapets, ornaments, and other projections on buildings other than private dwellings. The most common method for strengthening such elements is to reinforce them with steel ties and anchors and/or add the additional lateral supports of steel buttresses. Typically, enough time is allowed so that the problems can be eliminated in an orderly and thorough fashion and so that owners can finance the repairs. However, to date, most cities in California have not ensured that such repairs have been carried out.

(Figure 40)

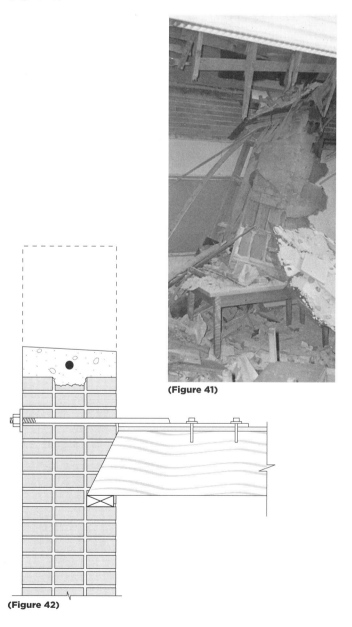
(Figure 41)

(Figure 41) During the 1971 San Fernando quake, a parapet collapsed and fell through the ceiling of a classroom at Los Angeles High School. After numerous repetitions of this kind of damage to schools, California state and local governments took steps to replace or reinforce unsafe schools.
(Figure 42) One relatively simple method of parapet reinforcement is shown here. The parapet is shortened, anchored with steel bolts and straps, and capped with concrete.

(Figure 42)

CHAPTER 7

Chimneys and Fireplaces

Exterior masonry chimneys are usually the most prone to damage and are often the most damaging element of the typical wood-frame house. In fact, seismologists used to determine the intensities of a given earthquake partially by figuring the percentage of destroyed chimneys per unit area. In the 1906 San Francisco earthquake, San Mateo lost 92 percent of its chimneys; Redwood City, 96 percent; Belmont and San Carlos, 88 percent; Burlingame and San Mateo Hills, 73 percent. In the 1971 San Fernando earthquake, most of the houses in the area were less than fifteen years old, and their chimneys had been reinforced with vertical steel bars in accordance with local building codes. As a result, 68 percent of the chimneys in the highest intensity areas survived the quake without damage. Certainly, if the chimneys had not been reinforced with steel, the percentage would have been close to zero.

Because a house and its masonry chimney are essentially separate and very different structures, they tend to respond to earthquake motions by pounding and pulling apart. Thus, chimneys must be tied thoroughly to the frame of the building, preferably with long steel straps that are embedded into the masonry and nailed to the joists of the various diaphragms of the building. Any masonry chimney predating about 1960 is unlikely to have adequate ties to the building and can be expected to collapse in an earthquake.

If your brick chimney is somewhere in the middle of the roof, does not extend up more than 3 or 4 feet, and breaks at the roof line in a quake, the damage is usually not serious and repairs will be relatively easy. More serious problems occur with chimneys on the side of the house. Newer chimneys, built during the past thirty years, are usually reinforced with steel. Such a chimney typically leans away from the house in a quake. If it breaks, it most often falls away from the house. Older chimneys sometimes break away from the house, too, but since they often lack any reinforcement—or their mortar is weakened—they disintegrate more easily. In fact, this mode of failure is less damaging to nearby adjacent houses.

(Figure 43) Tall masonry chimneys are probably the most dangerous feature of houses during earthquakes. Tall chimneys like this one fall apart even in moderate earthquakes.

(Figure 43)

(Figure 44)

(Figure 45)

(Figures 44 and 45) A collapsed chimney and the damage it caused to the roof and living room of an otherwise undamaged house in the M-6.7 1992 Big Bear (Los Angeles area) earthquake.

(Figure 46) This examples show a shorter chimney damaged by the 1971 San Fernando earthquake in which the falling old stone chimney destroyed a newer addition to the house.

(Figures 47 and 48) Anyone constructing a chimney and fireplace in earthquake country is wise to take advantage of the new and very safe prefabricated sheet-metal units. One such unit is shown in figure 47 for a single-story house. The strength, lightness, and flexibility of the metal eliminate all of the traditional deficiencies of masonry chimneys. The metal units are also less expensive and require less construction time. A finished chimney is shown in figure 48.

(Figure 46)

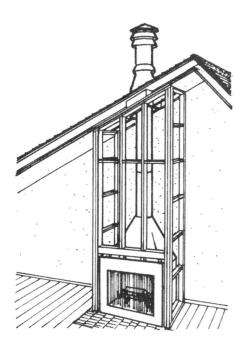

(Figure 47)

(Figure 48)

Tall chimneys in houses of two or more stories are much more susceptible to damage, and they can do more damage if they fall on another part of the house, or on a neighbor's. The farther a chimney rises above the roof line, the farther it can fall. If it extends more than about 5 feet, it may go through the roof if it falls toward the house, as is likely if the roof is flat or has only a gentle slope. If the roof is very steep, broken parts of the chimney tend to slide down without breaking through the roof.

If the chimney is tall and in the middle of the roof, 1-inch-thick structural-grade plywood can be laid either on the roof around the chimney (the preferred solution) or inside the attic around the chimney. The plywood should extend 1½ times the height of the chimney in each direction and should be nailed to the rafters.

It is difficult to reinforce an existing chimney properly without sizable expense. However, if your house has a very tall chimney, consider removing the portion above the roof and replacing it with a metal flue.

Brick chimneys over doorways are hazardous because they may fall on someone leaving the house in panic. Teach your family not to exit in a hurry through such a doorway.

Much better than the traditional brick and concrete chimneys are the prefabricated sheet-metal chimneys now available. These chimneys are very light, strong, and flexible and will likely not collapse or cause any pounding damage. For a traditional appearance, the sheet metal can be encased in a masonry veneer; far more economical and functional, however, is a well-designed covering of wood or stucco.

Problems with Old Buildings

The slow but steady evolution of building codes has led to a gradual improvement in the earthquake-resistant design of commercial and residential structures, particularly in California. This process was accelerated following the moderate but terribly destructive Long Beach earthquake of 1933. Before that time, some prominent geologists insisted that Los Angeles and its surrounding areas were in no danger from earthquakes, and building codes throughout Southern California reflected that attitude.

San Francisco was even more neglected. Soon after the scars of the destructive 1906 earthquake were removed, the business community of the city and its surrounding areas adopted the attitude that the fire, not the earthquake, had destroyed the city. Eastern financial interests and insurance companies were familiar with great urban fires, and San Francisco business leaders knew that, if the memories of the earthquake could be erased, much-needed funds for rebuilding could be more readily attracted. The disaster became known, then, as the great San Francisco fire, and local building codes regarding earthquake hazards were actually weakened in the years following the 1906 shock. Numerous large buildings built following the 1906 earthquake and during the speculative 1920s were designed to dangerously substandard requirements (although new research suggests that engineers and architects often designed beyond these code standards), and these are some of the buildings that will fail most spectacularly during the next large shock. As discussed in the previous chapter, the most dangerous of these are large concrete frame and masonry buildings, including both residential and commercial buildings. These are also among the buildings that suffered heavy damage in the 1989 Loma Prieta earthquake. A few came perilously close to collapsing.

Old buildings, no matter how elaborate or costly, can be a very risky investment, and old masonry and concrete buildings present the additional problem of considerable danger to their occupants. In addition, lack of adequate maintenance over the years has reduced the ability of many buildings to endure earthquake shocks. Rotted foundations or underpinnings, cracked mortar and plaster, and settled or distorted foundations, among many other symptoms of neglect, all undermine their earthquake resistance.

The minor Santa Rosa earthquake of 1969 provided some interesting data on the relationship between the age of a building and its susceptibility to quake damage. Of thirty-eight wood-frame dwellings badly damaged in the tremor, twenty-nine had been built before 1920 and the other nine predated 1940. No wood-frame structures built after 1940 sustained considerable damage.

If you own or have your heart set on a home built before about 1950, try to incorporate as many of the earthquake-resistant details and remedies described in this book into the house that you can. If you do, your new old home can be guaranteed a longer, more useful, and less costly and hazardous life.

(Figures 49-51) The old bungalow in figure 50 was one of several hundred wood-frame dwellings thrown off their foundations by the Long Beach tremor of 1933. The Victorian house in Santa Rosa (figure 51) also had foundation problems. The four-story, older building in figure 49 collapsed into a single-story pile of timbers, killing one person, in San Francisco's Marina district. It, like several of its collapsed neighbors, had a weak (soft) ground floor.

(Figures 52-54) While most pre-1950 homes in the Coalinga earthquake were severely damaged, as shown in figures 52 and 53, new single-story houses (figure 54) typically escaped significant structural damage.

(Figure 49)

(Figure 50)

(Figure 51)

(Figure 52)

(Figure 53)

(Figure 54)

WHAT DOES "TO CODE" MEAN?

Building codes are designed with public safety in mind. They are intended to make sure that buildings do not collapse, that people can leave buildings safely, and that hospitals and other critical facilities remain operational.

Building codes are not intended to eliminate potential damage to your property, to protect your investment, or to help your business keep going after the event. A building may perform to code standards—protect its occupants and allow them to leave the building safely (after the earthquake)—but still need to be demolished, causing you months of business downtime and large property and business-interruption losses.

Most of the buildings in the image on this page (showing damage following the 1995 Kobe earthquake), for example, performed "to code" in the sense that they did not collapse and that the occupants were allowed to escape the buildings after the earthquake. As an owner of a house, building, or business, you need to know the level of damage that your structure can sustain in an earthquake so that you can prepare for potentially large financial losses and business interruptions, and so that you can purchase earthquake insurance. If you don't take these precautions, you could save your life but lose your investment or business.

Structures can be designed to higher performance objectives than those required by building codes. So-called performance-based design, which in practice is typically used only for larger buildings, provides a means to make design decisions focused not only on life safety, but also on damage reduction and business continuity related to earthquake exposure. Performance-based design enables you to treat each building in light of its unique qualities and to meet your own particular needs.

The performance-based design procedure begins with understanding your building's level of earthquake risk and how you want your structure to perform after an earthquake. If you own or use an office building, you may want your building to simply remain standing after a major earthquake, allowing your staff to leave safely. However, if you operate a manufacturing facility, and any downtime would lead to lost market share, you may have a very different performance objective.

Of course, with increased performance, upfront design and construction costs may also increase. But this is usually a small expense when compared to the potential financial loss following an earthquake. Also, a higher design level should reduce the expected damage to a building, which, if handled properly by the owner and the insurance broker, should reduce the overall insurance premium. That, alone, may offset the higher upfront cost in a few years.

The advantage of the performance-based design approach is that you tell the earthquake what the damage should be, as opposed to the other way around.

Kobe, Japan 1995

THE LOW-PROBABILITY, HIGH-CONSEQUENCE BIG EARTHQUAKE

Building codes take into account not only the consequences of an earthquake but also the probability that the quake will occur. The required strength of the buildings (or the design forces in the codes) is reduced if the likelihood (not the consequences) of that earthquake occurring is smaller. The result is that buildings in areas such as New Madrid, Missouri (including St. Louis)—the location of the most powerful earthquake in U.S. history (1812)—are designed for significantly lower forces than in California. The same is the case in the Portland, Seattle, and Vancouver areas. Because of that, should a major quake occur in these regions, the consequences will be devastating. In effect, all new buildings, including numerous high-rises in all of the above cities, are designed for much smaller earthquakes than they may ultimately experience. That will result in far more damage than is expected in California, where design requirements are more realistic simply because large quakes happen more frequently. We believe that this inadequacy may result in spectacular collapses of numerous large, modern buildings such as office towers or high-rise apartment buildings. This issue has not received adequate attention from either the engineering profession or city planners. Meanwhile buildings, especially condominium buildings, are getting taller and pushing the design envelope as discussed in the sidebar on page 95.

Seattle, Washington

St Louis, Missouri

RETROFIT AND NEW DESIGN OF COMMERCIAL BUILDINGS FOR EARTHQUAKE LOADS

Many advances in earthquake science, engineering, and technology have taken place over the past few decades. These advances are slowly being incorporated into existing structures (through retrofit design). These advances are also applied to new buildings throughout the world, allowing for architectural visions to come to reality, while creating structures that protect life, property, and business.

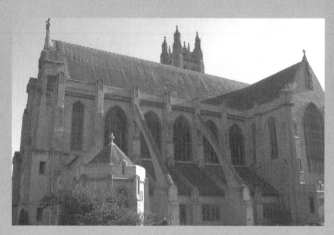

Flying buttresseses were added in 2007 to strengthen this old concrete church in San Francisco. Strengthening members can be hidden or exposed, depending on the structural and architectural needs of the project.

This old reinforced and infill brick building in Sacramento, California, was recently strengthened using an innovative approach using diagonal steel braces with energy absorption mechanisms called friction dampers.

This pre-cast concrete building on the UC Berkeley campus was retrofitted using innovative unbonded braces. Unbonded braces are made of steel (which is strong in tension) and concrete (which is strong in compression), enabling the brace to possess similar properties in both tension and compression.

The world's only "rocking" building: Conventional structural design for this fifteen-story Maison Hermès building in central Tokyo's Ginza district resulted in a high weight of steel in the frame, high foundation forces, and excessive construction costs. Inspired by traditional Japanese pagodas and temples, its innovative design includes a system by which the columns at one side of the narrow building are allowed to lift up rather than develop forces at the foundation. The uplift is controlled by dampers (above photo) using the same concept as shock absorbers in your car. This greatly reduces the earthquake forces in the structure and the foundation and saved the project millions of dollars in project costs.

BASE ISOLATION

Base isolation is another innovative form of earthquake protection. There are two schools of thought with earthquake resistant design: One is to compete against the earthquake, and the other is to not play its game.

In the first case, the structure is designed to be strong enough to resist earthquake forces. However, for economic and aesthetic reasons, almost all structures are designed to sustain some damage in major earthquakes, thus limiting the amount of force and acceleration transmitted through the structure. Hopefully this damage is sustained in members that do not support the structure against gravity (like columns).

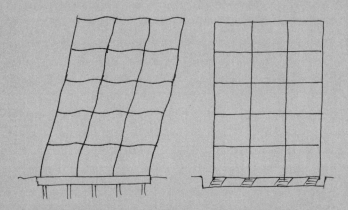

In the second case, the designer separates, or isolates, the structure from the earthquake, usually at the base of the structure, and is hence called "base isolation." See schematic to the right.

This is achieved through several means, including sliding and rubber bearings (as shown being tested in the bottom photo). Contrary to myth, buildings are not placed on rollers. The bearings include some means of recentering the structure after the earthquake, and some means of limiting displacements across the isolator through energy absorption or "damping."

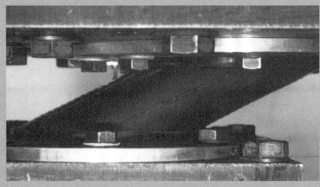

The great thing about base isolation is that displacements and accelerations within the building are greatly reduced resulting in less structure to resist the earthquake forces, and substantially less direct and indirect financial loss. In most cases, the estimated financial loss is less than insurance deductibles.

Japan has embraced this technology to a greater degree than the United States, but its use is gradually becoming more widespread.

If you are a building owner, discuss base isolation with your structural engineer. It is a good way to retrofit existing structures, as well as protect critical new architecture (like medical facilities, data centers, and buildings that house sensitive equipment). It could save you real money in the short and long term. Two examples of structures that were retrofitted using base isolation are shown on the facing page.

Hearst Memorial Mining Building, Berkeley, California

City and County Building, Salt Lake City, Utah

CHAPTER 7

San Fernando, California, 1971

Niigata, Japan, 2007

CHAPTER 8

How to Minimize Interior, Utility, and Equipment Damage

Some interior damage is inevitable in a strong earthquake. Dishes will fall and break. Tall furniture will topple, and pipes may rupture. Generally, this type of damage can only be anticipated; to try to prevent it entirely would be to turn your household into a museum (or a motel room) in which everything is securely locked away or attached to the walls and floor. However, by anticipating the effects of an earthquake, you can minimize such damage and, in some cases, select the household items that you wish to protect above all others. This chapter covers a miscellany of objects, utility components, and interior features that are most typically subject to earthquake damage but can be protected to some extent.

If you own a business, you should know that nonstructural damage, especially damage to equipment, is often a major cause of expensive business interruptions. We briefly summarize the types of damage that will occur if the necessary strengthening features have not been added.

Utilities

The rupture of water piping during an earthquake can cause extensive interior water damage. Fortunately, this is not normally a problem, because waterlines are broken only when serious structural damage occurs. Gas mains, however, should be thoroughly checked immediately following a strong quake. Keep in mind that leaking gas lines can cause a fire well after the shaking stops. This was the cause of the San Francisco Marina fires following the 1989 earthquake and the many fires that burned numerous properties in the San Fernando Valley shortly after the Northridge earthquake of 1994. Several blocks of Kobe, Japan, burned for the same reason in 1995. If you smell gas and suspect or discover a broken gas line or gas leakage, shut off the main feeder valve and do not turn it on again without the assistance of a utility company employee. Be aware that the utility company employees might be very busy and will not be there for a very long time. You should therefore be prepared to live without utilities for quite awhile.

If severe ground settlements or slides occur in the vicinity of your property during an earthquake, the gas lines are very likely to be damaged. You will know there is such damage, of course, if the gas is not reaching your home. In this situation, the best course of action is (1) if possible, notify the utility company immediately; (2) prepare warning signs around the likely gas line breakage to keep other people away from the danger; and (3) move your family and valuables a safe distance from the probable break.

To protect against utility damage, have a plumber evaluate, replace, and properly secure rusted or worn water and gas pipes. The plumber can also replace rigid gas connections to water heaters, stoves, dryers, and other gas appliances with flexible stainless-steel gas connectors. Excess-flow gas-shutoff valves for individual appliances, which stop gas flow in case of a leak, are now available for use with flexible connectors.

(Figure 1)

(Figure 2)

(Figure 3)

(Figure 4)

(Figure 5)

(Figure 1) Electric power substations and other distribution nodes of the power grid tend to suffer severe damage, even in smaller earthquakes, because many of their porcelain components are weak. This damage occurred near Monterey in 1989.

(Figure 2) Fire following an earthquake is a very real possibility when structural failures in a building, particularly in the foundation area, are likely to damage and sever gas pipes. This photograph shows the ruptured gas line under the porch of a house that fell off its foundations in Santa Rosa in 1969.

(Figure 3) The same damage occurred to many houses in Watsonville, near San Francisco in October of 1989. Note the broken gas meter on the house (which also fell off its foundation).

(Figure 4) A neighboring house was not as lucky and burned down from a gas ignition—all that is left are the front steps.

(Figure 5) During the 1989 earthquake, settlement of poor ground in San Francisco's Marina district severed numerous gas and waterlines, incapacitating the fire protection system.

FIRE FOLLOWING AN EARTHQUAKE

Earthquakes in urban areas are often followed by destructive fires caused by gas line breaks and electrical shorts. The situation is often exacerbated by damaged water tanks and broken water pipes that restrict firefighter access to water for firefighting, and damaged infrastructure. Fires following earthquakes can cause extensive damage, often burning entire neighborhoods (as seen in the image on the right after the 1994 Northridge earthquake), or in the case of the 1906 earthquake, an entire city (below).

Los Angeles, 1994

San Francisco, 1906

BE PREPARED TO SHUT OFF UTILITIES (BUT ONLY WHEN NECESSARY)

Gas Shutoff

1. Locate the main gas shutoff (usually outside the house) and all pilot lights.
2. Clear the area around the shutoff valve for quick and easy access in case of emergency.
3. A wrench (or specialty tool for turning off gas and water) should be attached to a pipe next to the shutoff valve or in another easily accessible but hidden location.
4. You may want to paint the shutoff valve with white or fluorescent paint so that it can be located easily in an emergency.
5. If you are concerned about your ability to turn off the main gas shutoff valve, are unsure if it is in proper working order (you see indications of rust, for example), or do not know how to relight your pilot lights, contact your local gas company. It can send a service representative to show you the proper procedure and check the valve and pilot lights to be sure they operate properly.

Water Shutoff

1. Locate the main water service pipe into your house (probably in the front at the basement level). You will see a gate valve on the pipe. If you know you have leaks after an earthquake, you can shut off all water in your house with this valve. You may wish to paint the valve so it is easy to find in an emergency.
2. You can shut off all water to your property by finding the water meter box (usually at the street or sidewalk). Open the cover with a long screwdriver or specialty tool. The shutoff valve can be operated by a specially designed tool or a crescent wrench. If this box is inaccessible or you cannot find it, call your local water department.

Electrical Shutoff

1. Locate the main electrical shutoff.
2. Your house may be equipped with fuses or circuit breakers. If your house has fuses, you will find a knife switch handle that should be marked "MAIN." If your house has circuit breakers, you may need to open the metal door of the breaker box to reveal the breakers (never remove the metal cover). The main circuit breakers should be clearly marked with their on and off positions. If there are any sub-panels adjacent to the main fuse or breaker panel or in other parts of the house, shut them off, too.

Note: All responsible family members should be shown how to turn off utilities in emergencies.

Water Meter

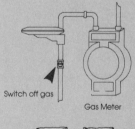
Gas Meter

Special Tool available at some hardware stores for shutting off gas and water (A crescent wrench will work.)

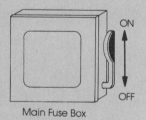
Main Fuse Box

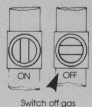
Switch off gas

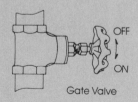
Gate Valve

Turn off main toggle only

Simplified Circuit Breaker Box

STRAPPING YOUR WATER HEATER

Strapping your water heater and making sure it is fitted with a flexible gas supply line will greatly reduce the danger of fire or explosion from a gas leak after an earthquake. If your water heater does not have a flexible gas supply line, contact a licensed plumber to install one. The instructions below are for a 30- to 40-gallon water heater within 12 inches of a stud wall.

1. Mark the water heater at 6 inches down from the top and about 18 inches up from the bottom. Locate the wood studs in the wall on both sides of the water heater. Transfer the marks on the heater to the wall on both sides.
2. Drill a 3/16-inch hole 3 inches deep through the wall sheathing and into the center of the studs at the four marks made in step 1.
3. Measure a circle around the water heater and add 2 inches to the measurement. Using a hacksaw, cut two 1½-by-16-inch-gauge metal straps to this length to encompass the water heater.
4. Mark 1½ inches from each end of the straps and insert in a vise or under a heavy object. Bend the ends outward to approximately a right angle. Bend the rest of the straps into a curve.
5. Measure the distance from a point midway on each side of the water heater to the holes drilled in the wall. Add 1½ inches to these measurements. Using a hacksaw, cut two pieces of conduit to each of these two lengths.
6. Flatten approximately 1½ inches at each end of the four pieces of tubing by laying the tubing on a flat metal or concrete surface and striking it with a hammer. Be sure you flatten both ends on the same plane.
7. Insert the flattened ends of the tubes, one at a time, into a vise or clamp. With the hammer and a center punch, make a mark ¾ inch from each end at the center of the flattened area of the tube. Drill 3/8-inch holes in the ends of all four tubes (making eight holes in total). Be sure the tubes are clamped down while drilling. Bend each end to about 45 degrees.
8. Wrap the straps around the heater and insert a 5/16-by-1¼-inch bolt with washers into the bent ends. Tighten the nuts by hand. Insert 5/16-by-3/4-inch bolts through the strap from the inside at the midpoint on each side of the water heater. Attach one end of each tube strut to a protruding bolt, add a washer and nut, and tighten by hand. Insert a 5/16-inch lag screw in the opposite end of each tube strut and insert in the hole in the wall stud. You may need to tap the lag screw gently into the hole, and then tighten it with a crescent wrench.
9. Adjust the straps to the proper height and tighten all nuts snugly, but do not overtighten.

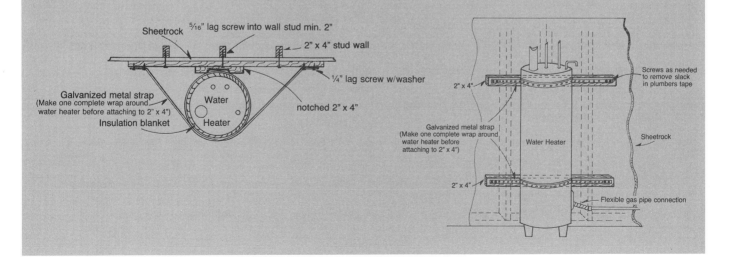

Water Heaters

Of all household utility components, unbraced water heaters are most vulnerable to earthquake damage and can be a major fire hazard. Because they are quite heavy and often stand on supports unconnected to the floor or walls, they readily topple over, tearing out rigid gas lines and allowing gas to escape freely. See that your water heater, gas or electric, is bolted to the floor (many models provide bolt holes for this purpose) or (preferably) strapped securely to the adjoining walls or structure. Flexible gas supply lines should also be installed.

Kits for securing your water heater are usually available at major hardware stores. (See "Strapping Your Water Heater" on page 149.) The entire operation will probably cost under $25, and those twenty-five bucks will buy you a lot of peace of mind.

Furnaces

Household furnaces of all types should also be anchored; otherwise they can slide and rupture gas lines. Typically, they are even easier to secure with a few bolts or straps than water heaters are.

Air Conditioners

Give some thought, too, to the location and installation of air conditioners. For example, it is not advisable to place them in the vicinity of a masonry chimney, which may collapse during a quake. The best location for an air conditioner is at ground level along the side of a building, where it is anchored to a low concrete mat foundation.

Plaster and Wallboard

Heavy plaster finishes in old houses, particularly plaster ceilings, can pose a considerable hazard to furnishings and occupants. Old plaster ceilings, for example, may weigh as much as 8 pounds per square foot, which means that the ceiling of a small 15-by-15-foot room could weigh a ton or more. On the other hand, good-quality plaster, such as exterior wall stucco, is surprisingly strong and will crack only under very heavy shaking and large wall deformations. Indeed, the combination of lath and thick plaster has a stiffening effect that can significantly increase the overall earthquake resistance of the walls.

Falling or cracking plaster can best be avoided by ensuring that the plaster is in good condition (without long cracks or spots softened by water leakage). If you own or will purchase an older house with plaster on inherently weak walls, such as non-bearing partition walls, you should anticipate plaster damage in an earthquake.

Fortunately, wallboard, also known as gypsum board and sheetrock, has replaced plaster in almost all construction since the early 1950s. If a house has been properly stiffened and strengthened with plywood shear walls, or it is single-story, then its wallboard will not be easily damaged in an earthquake. In a multistory house, however, wallboard can be easily damaged if the house does not have shear-wall bracing. The good news, though, is that damage to wallboard is usually not dangerous to occupants and is easily repairable.

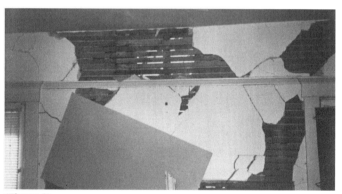

(Figure 6)

(Figure 6) Plaster, particularly cracked plaster in older homes, applied over lath backing, is more easily damaged than Sheetrock. This house in Whittier, in Southern California, lost much of its interior plaster in the 1987 earthquake.
(Figures 7-9) Interior damage can be very severe when houses fall from their foundations or when soft-story buildings partially collapse. These examples are from the Marina district in 1989.

(Figure 7)

(Figure 8)

(Figure 9)

Furnishings

A strong earthquake is bound to cause some damage to furnishings, particularly breakable items such as china, glassware, pictures and picture frames, lamps, and art objects. But much of this damage can be prevented if you are willing to take the time.

If you don't want to bother with all your decorative objects, at least consider protecting (1) valuable paintings that might tear if they fall on furniture and (2) objects that might gouge expensive furnishings. Paintings and other hung art objects should be attached to the wall with oversized threaded hooks, which can be screwed deep into the studs. For small frames, use regular but oversized hangers, and close the open hook to make it difficult for the wire to slip out during an earthquake.

Falling bookcases and books are common in earthquakes and can pose a hazard to occupants. Tall bookcases and other tall furniture with narrow bases easily fall. It is wise to attach such heavy furniture to wall studs with small steel angles or straps. Major items of furniture purchased in earthquake country often come equipped with such angles or straps, saving the separate purchase of these items. Flexible-mount fasteners, such as nylon straps, allow furniture independent movement from the wall, reducing strain on studs. It's also a good idea to arrange heavy objects on the lower shelves of a bookcase or cabinet. A lightweight bar, dowel, or rim across the front of shelves is another protection for dishes and other objects displayed in high places.

Most damage to china and glassware occurs when cabinet doors fly open and the contents fall to the floor. The easiest and best solution is simply to place strong push latches on your cabinets. Whatever type of latch you use, the point is to keep the doors closed against the force of the contents during an earthquake. Some hardware stores sell a special type of latch that opens only after a cabinet door has been pushed from the outside. Childproof or boat-safety latches can also be effective.

Valuable glass or other fragile objects can be held in place with removable earthquake putty, museum wax, quake gel, or discreet wiring. When they are not displayed, always keep these items in locked drawers or cabinets, preferably in a low-profile piece of furniture. Low cabinets are far less likely to overturn during a large tremor.

One more note: It is extremely foolhardy to place beds anywhere near high, heavy objects—for example, a chandelier, a stereo, tall heavy furniture, or a large mirror. The first jolt of a strong earthquake may awaken you—but it may also plunge that heavy object into your bed before you can move to safety. It's also wise to place beds away from large glass walls or windows unless they have heavy curtains or blinds to protect you from falling glass.

(Figure 10)

(Figure 11)

(Figure 12)

Swimming Pools

The hazard from swimming pools, especially those on hillside sites, is possible damage from sloshing and escaping water. In large California quakes, pools have often lost as much as 25 percent of their water from earthquake motion. Always locate pools as far as possible from the house and at a lower elevation.

Freestanding and Retaining Masonry Walls

Masonry retaining walls and fences are heavy, inflexible, and segmented, and may lack support (such as freestanding garden walls). They can easily collapse without good footings, reinforcement, and full grouting. Collapse is not generally a hazard to life or property, of course, but the replacement of these walls is expensive, and the damage is easily avoided. Consult an engineer as to the best way to reinforce these walls

(Figure 13)

(Figure 10) Some interior damage and much disarray of objects and furnishings are inevitable in a large earthquake. But such damage—seen here in a home in the San Fernando Valley—can be minimized if you anticipate the effects of a quake and secure your most breakable and valuable belongings.

(Figures 11 and 12) Views of interior damage to a hillside house after the Morgan Hill earthquake of 1984. The house itself had no significant structural or exterior damage.

(Figure 13) Concrete-block retaining and garden walls call for the same type of careful reinforcement that is required in constructing a concrete-block building. All of the garden walls on this block in the San Fernando Valley were toppled by the 1971 earthquake. Note that the columns remained intact when they fell because they had been grouted with concrete. Had they also been reinforced with a steel rod that extended into the base, they and the whole wall would have remained upright.

REDUCING INTERIOR DAMAGE

Cabinets
Install positive catching latches. Many variations designed for earthquake safety are available at hardware stores.

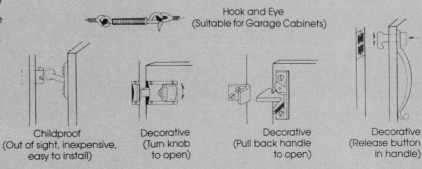

Tall Furniture
In addition to the angle brackets shown here, you may want to also install flexible fasteners, such as nylon straps, which allow furniture to move independently from the wall, reducing strain on the wall connection. Several of these fastener types should be available at your hardware or home-improvement store. Make sure that the fasteners are attached securely to wall studs.

Open Shelves
Install a guard across the shelf, or install wood trim on the front of the shelf. Place heavy objects on lower shelves.

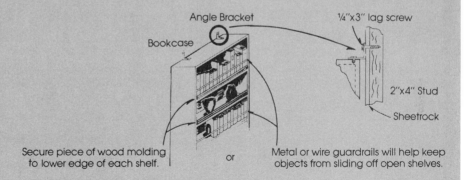

Hanging Pictures
Screw hooks into wood members only (stud or ceiling joints). Close hooks used for hanging pictures to prevent them from falling.

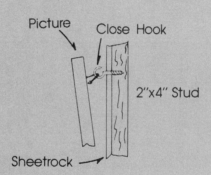

AVOIDING BUSINESS INTERRUPTION

Business interruption associated with facility and equipment downtime can cause significant financial loss. Following are several images of damages that have caused significant business interruptions in previous earthquakes. As well as considering direct physical damage to your facility, you must also consider the ability for your staff to get to work as well as your reliance on public utilities that might not be available. Your Business Continuity Plan (see next chapter) should take these considerations into account.

INTERIOR DAMAGE

(before)

(after)

DAMAGED CEILINGS

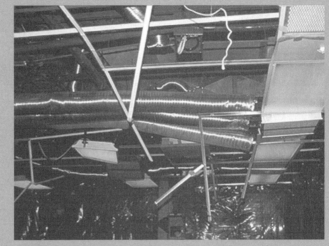

COLLAPSED RACKS

CHAPTER 8

DAMAGED EQUIPMENT

If there is one lesson to learn from the 2007 Niigata, Japan, earthquake, it is that equipment, like buildings, needs to be protected from earthquakes. Note that this electrical control cabinet (left) was properly anchored with bolts to the floor and remained functional while the structurally weak building collapsed around it. Damage to equipment that was not anchored caused most of the loss in this earthquake (several billion U.S. dollars)—primarily due to business interruptions. Most equipment in industry is not properly anchored. The needed anchorage or bracing can be simply designed and installed

DAMAGED COOLING TOWERS AND TANKS

FIRE FOLLOWING

TRANSPORTATION SYSTEM FAILURE

PROTECTING YOUR WINE

In a strong earthquake, objects, and particularly objects placed on shelves, move and fall off. As shown in the sidebar on page 156, objects can move or slide several feet in strong ground motion. It is always sad to see a winery affected by an earthquake. Thousands of broken bottles and toppled and cracked barrels, however, create a wonderful aroma—at least for a few days.

Most wineries have a serious earthquake problem, mainly due to lack of adequate protection of their wine inventory. That includes the way they store their ageing barrels (right), and more important, the stainless-steel tanks that contain the new wine (bottom right). Stainless steel tanks need to be properly anchored to their concrete bases. Also, because the tanks are very thin, they tend to be damaged severely when full. A structural engineer is needed to provide the necessary design.

If you have a wine cellar, or designated shelving where you store your wine or wine collection, you should take a few simple steps to protect the bottles from flying off the shelves and breaking. The easiest way, as illustrated in the top photo on the facing page, is to provide sliding doors in front of the bottles that would stop the sliding bottles from falling out. The doors can be standard sliding doors, or, if you prefer to see the stored bottles, glass doors. The glass should be tempered. Such doors should prevent most of the potential damage, but not all of it, as the doors may slide somewhat. Childproof-type locks should eliminate almost all of the damage, if you want to bother with them. If the wine is stored in boxes, provide stops in front of the boxes so that they cannot slide off.

An appropriate means of storing barrels is shown in the bottom photo of the facing page. Note the steel brace protecting the winery's storage of barrels.

Santa Cruz Mountains, California, 1989

Los Angeles, California, 1994

CHAPTER 9

Earthquake Insurance

Is earthquake insurance available and affordable? Do you need it? And if so, how much and what kind? Although your answers to these questions will be based on your individual situation and your earthquake risk-management strategy (see the introduction on page 12), this chapter provides you with an overview of earthquake insurance and whether it may be appropriate for your unique needs.

For homeowners, the purchase of earthquake insurance is often an emotional decision. Typically fewer and fewer homeowners carry the insurance as memories of the last destructive quake fade. Most have forgotten the lessons of the last really damaging and now-distant 1994 Northridge earthquake. Insurance has also become quite expensive—it typically doubles a homeowner's annual premiums.

For business owners, the cost and availability of earthquake insurance changes from year to year. This often results in many corporations (some very large corporations) leaving themselves unnecessarily exposed to earthquake loss. Although this book does not address the complex issue of commercial earthquake property insurance through our Earthquake Risk Management Program, we recommend an approach that balances insurance with other means of managing risk, including risk mitigation and risk retention. The result is a stabilization of earthquake risk management costs from year to year.

An Introduction to Earthquake Insurance

Earthquake insurance is a form of property insurance. The policyholder pays a premium for protection in case an earthquake causes property damage. In general, in the United States, most ordinary homeowners' insurance policies do not automatically cover earthquake damage. This is not the case in California, where earthquake insurance must be offered in conjunction with an all-risk insurance policy. (This is similar to the situation in Florida, where hurricane insurance takes the place of earthquake insurance.)

With any catastrophe affecting an entire region, insurance companies need to be very careful when providing coverage. If an earthquake is strong enough to destroy one home, it will probably damage or destroy hundreds of other homes in the same area. If one company has written insurance policies for a large number of homes in that region, then claims could quickly exceed the company's financial reserves. This happened to 20th Century Insurance, which was based in the San Fernando Valley and was bankrupted by the 1994 earthquake.

Like other catastrophe risk policies, most earthquake insurance policies feature a high deductible, which makes the insurance useful if a home is severely damaged or destroyed, but irrelevant if a home is damaged below deductible limits. Deductibles can range anywhere from 2 percent to 20 percent of the replacement value of the structure. This means that if the replacement cost of your home is $500,000 and your policy has a 10 percent deductible, you would be responsible for the first $50,000 of damage. In most cases, consumers can select higher deductibles for reduced premiums. Insurers in states such as California, Washington, Nevada, and Utah, with higher-than-average risk of earthquakes, often set minimum deductibles at around 10 percent; 15 percent is the standard deductible in California.

For insurance purposes, an earthquake is defined as any one shock (or a series of several shocks) that occurs in a given seventy-two-hour period. The deductible may be applied more than once for damaging shocks that occur at intervals greater than seventy-two hours. Earthquakes or aftershocks that occur after seventy-two hours have passed are considered to be separate earthquakes, therefore the deductible would apply again.

Premium rates depend mostly on the likelihood of a damaging earthquake in your region. Typically, rates in California are higher than they are in Utah, for example, because the probability of a damaging quake is greater. Premiums can also differ by location (and soil type), age and type of construction, and the insurance company itself. Generally, older buildings cost more to insure than new ones. Wood-frame structures generally benefit from lower rates because they tend to withstand quakes better than other forms of construction do.

The cost of earthquake insurance is calculated on a "per $1,000 basis." The premium for a house in the Pacific Northwest, for example, may cost $1 to $3 per year per $1,000 worth of coverage, while it may cost less than 50 cents per $1,000 on the East Coast. Premiums in California can be substantially higher, $3 to $4 per year per $1,000 worth of coverage.

Premiums and deductibles are determined by the replacement value of the primary building or house before the loss. The land and "appurtenant" structures such as detached garages or swimming pools are sometimes included, too, outside California. Your coverage must reflect the current costs for rebuilding the primary (dwelling) structure. An undervalued property could result in a lower deductible but also could be completely unrealistic: you might not receive enough money to replace or repair a badly damaged structure. Remember, the cost of a major rebuilding project is financially devastating to most homeowners. The goal of earthquake insurance is to insure against this event.

Window and door glass is generally not covered against earthquake damage. If such coverage is desired, it must be obtained in a separate policy.

Loss caused directly or indirectly by explosions, floods, or tsunamis resulting from an earthquake is not always covered by earthquake policies. Be certain that you understand the range of coverage, and consider the various earthquake risks to your property. If you live near the ocean shore in a sea-level site in Alaska, Washington, Oregon, and parts of California for example, tsunami coverage may be missing.

Unfortunately, people have short memories when it comes to catastrophes, especially earthquakes. In general, too few carry adequate insurance. Only about 12 percent of Californians purchased earthquake coverage in 2007 (or about 760,000 policies), down from about 30 percent in 1996, when the destruction of the 1994 Northridge earthquake was still fresh in people's minds. This runs against common sense, as the probability of an earthquake increases with time.

Earthquake Insurance in California and the California Earthquake Authority

A legal theory known as concurrent causation holds that, under a so-called all-risk policy, such as the typical homeowners' insurance policy, if a loss is attributable to two perils (for example, fire following an earthquake), and one of which is excluded while the other is not, the loss is covered. In the past, earthquake damage may have been paid under policies without specific earthquake coverage if a nonexcluded contributing cause, such as negligent construction, could be found to have concurrently caused the damage. Therefore, the pricing of normal homeowners' policies used to include a crossover component for earthquake risk coverage even though the stated terms of the homeowners' insurance policy explicitly excluded earthquake coverage.

The California legislature enacted a law, effective January 1, 1985, to alleviate problems caused by the concurrent causation theory. Their intent in enacting Assembly Bill Number 2865 was "to make clear that loss by or resulting from an earthquake shall be compensable by insurance coverage only when earthquake protection is provided through a policy provision or endorsement designed specifically to indemnify against the risk of earthquake loss, and not through policies where the peril of earthquake is specifically excluded even though another cause of loss acts together with an earthquake to produce the loss." In short, the legislation made clear that if you want to be protected in the next earthquake, you better have an explicit earthquake insurance policy. It is important to note that this legislation also required that insurers offering all-risk policies to homeowners must also offer earthquake insurance.

Then came the 1989 Loma Prieta earthquake. For most of the period since the great 1906 San Francisco quake, earthquakes had been profitable for the insurance industry, with claims and payments being far less than premiums, even including three damaging California earthquakes (1971 San Fernando, 1983 Coalinga, and 1987 Whittier Narrows). But payments rose dramatically after Loma Prieta, from less than $3 million for the 1971 San Fernando earthquake to more than $1 billion for the 1989 quake.

If the 1989 quake shook the insurance industry, the 1994 Northridge earthquake—with insured residential damage ultimately totaling $12.5 billion—brought it to its knees. The situation was made worse still by the $200 billion in losses in the Kobe, Japan, region from the January 1995 earthquake there. The insurance companies' response? By mid-1995, insurers representing about 93 percent of the homeowners' insurance market in California severely restricted—or refused to write—new homeowners' policies because of the mandate that they also offer earthquake insurance. The financial implications for the state were obvious.

One remedy considered by legislators was to repeal the law that required earthquake coverage to be provided to all-risk policyholders—but that answer would have left millions of homeowners unprotected, and ultimately it was rejected. To solve the crisis, the State of California designed a catastrophic residential earthquake insurance policy that eventually led to the formation of the California Earthquake Authority (CEA) in September 1996. Today, the CEA is the world's largest residential earthquake insurer, providing coverage to California homeowners, renters, condominium owners, and mobile-home owners through its participating insurance companies. Companies that sell over two-thirds of the residential property insurance in the state have opted to offer the CEA policy.

At the time this was written, the CEA had access to over $7.2 billion for the payment of potential claims. If an earthquake causes insured damage greater than that, then policyholders who suffered damage will be paid a prorated portion of their losses. The resources available to pay claims should grow over time, especially if Californians are lucky enough not to experience really damaging earthquakes for a few more years.

As stated earlier, insurers that sell residential property insurance in California must offer their policyholders earthquake insurance. Companies can offer their own private earthquake policy or they can offer policies through the CEA. Most policies offered by CEA companies are similar to the CEA policies. CEA policies have some limitations as noted on pages 164-165.

California Earthquake Authority Policies

The CEA homeowners' policy pays to repair an insured dwelling when loss from covered earthquake damage exceeds the deductible. Pools, spas, fences, patios, and many other features are not covered; detached garages and other outbuildings are also excluded from coverage. The table in the next page provides an overview of the type of protection provided by the CEA.

Should You Buy Insurance?

You should buy earthquake insurance only if it makes financial sense—in other words, if you need it. Too often, people (and some of the world's largest corporations) make earthquake insurance decisions without considering their overall risk management strategy. How to balance the various means of managing risk in earthquake country is discussed in detail in the introduction. But our recommendation is simple: it is easiest, and likely most cost effective, to hire a structural engineer and ask him or her to orally estimate the damage that you might suffer (after all life-threatening features have been retrofitted). If the expected non-life-threatening damage is significantly higher than the deductible, or if earthquake insurance gives you additional peace of mind, then buy the insurance.

Neither of the authors of this book carries earthquake insurance on his primary residence. We have retrofitted what needed to be strengthened and analyzed the expected damage, and we do not foresee that expected damage would significantly exceed the deductibles. We may be surprised, but the probability of that is low.

We do stress, however, that earthquake insurance will not protect your family from injuries. Making sure that your family members are safe in the next earthquake should be your highest priority.

CALIFORNIA EARTHQUAKE AUTHORITY POLICY INFORMATION*

COVERAGE		HOMEOWNERS	MOBILE HOME
A: Dwelling	Limit	Insured value of your home	Insured value of your home
	Deductible	Policyholders may select either a 10 percent or 15 percent deductible. Only structural damage counts toward meeting the deductible. Unless a CEA-insured homeowner suffers damage to the structure exceeding the amount of the deductible, the CEA will not pay any claimed loss for structure or contents	Policyholders may select either a 10 percent or 15 percent deductible. Only structural damage counts toward meeting the deductible. Unless a CEA-insured homeowner suffers damage to the structure exceeding the amount of the deductible, the CEA will not pay any claimed loss for structure or contents.
B: Extensions of dwelling	Limit	Limited coverage for utility structures, ingress and egress, and retaining walls.	N/A
	Deductible	(see above)	N/A
C: Contents	Limit	Up to $5,000 to repair or replace personal property, such as furniture and appliances.	Up to $5,000 to repair or replace personal property, such as furniture and appliances.
	Deductible	See above.	See above.
D: Loss of use	Limit	Up to $1,500 for additional living expenses you might incur if your house is uninhabitable after an earthquake.	Up to $1,500 for additional living expenses you might incur if your house is uninhabitable after an earthquake.
	Deductible	None	None
E: Loss assessment	Limit	N/A	N/A
	Deductible		

CONDOMINIUM	RENTERS
Up to $25,000 to repair or replace structural components of your condominium for which you are legally responsible.	N/A
$3,750 (15 percent of $25,000 limit of insurance)	N/A
N/A	N/A
N/A	N/A
Up to $5,000 to repair or replace personal property, such as furniture and appliances.	Up to $5,000 to repair or replace personal property, such as furniture and appliances.
$750	$750
Up to $1,500 for additional living expenses you might incur if your house is uninhabitable after an earthquake.	Up to $1,500 for additional living expenses you might incur if your house is uninhabitable after an earthquake.
None	None
Pays for certain earthquake damage assessments made against you by your homeowners' association. If the market value of your condominium is greater than $135,000, you can purchase $50,000 in coverage; if the market value is $135,000 or less, you can purchase either $25,000 or $50,000 in coverage.	N/A
15 percent of the limit of insurance.	

Certain supplemental coverage limits are available. If you are a homeowner or mobile-home owner, for an additional premium you can lower your deductible to 10 percent instead of the standard 15 percent. All CEA policyholders can raise contents coverage as high as $100,000 and loss of use up to $15,000.

If you are a homeowner, your house may be eligible for a 5% premium retrofit discount if it was built before 1979 and its sill plate has been bolted to its foundation, its crawlspace walls have been braced, and its water heater has been strapped (all of which you can do through guidance in this book). No retrofit discount is available for houses built on concrete slab foundations.

*Note: This table is for general information only and was valid at the time of writing. You should consult the CEA Web site http://www.earthquakeauthority.com for more and up-to-date information.

Niigata, Japan, 2007

CHAPTER 10

What to Do Before, During, and After an Earthquake

Before an Earthquake

The most important measures in preparing for an earthquake are (1) making your building and its contents stable and earthquake resistant for life safety; (2) finding a balanced approach to strengthening and insurance; and (3) thinking about and discussing with your family, neighbors, and coworkers the likelihood and effects of the next big earthquake, and considering what you and the others would and should do in your home, at work, in your car, or in any other place when the tremor strikes. Conscious preplanning will enable you to react calmly and effectively during the emergency.

During an Earthquake

Your most important task is to remain calm. If you can do so, you are less likely to be injured, and those around you will also benefit from your coolness. Think about possible consequences before making any moves or taking any actions.

In the United States, the vast majority of injuries during an earthquake are due to falling objects and debris, not collapsed buildings. If you are inside a building, get under a sturdy desk or table. Watch for falling plaster, suspended ceilings, light fixtures, and tall, heavy furniture. Stay away from windows, large mirrors, and masonry chimneys and fireplaces. In office buildings, it is best to get away from windows or glass partitions, and watch out for falling ceiling debris. In factories, stay clear of heavy machinery and equipment that may topple or slide across the floor.

Do not rush outside if you are in a store, office building, auditorium, or factory. Stairways and exits may be broken or littered with debris and are likely to become jammed with panicky people. In addition, many deaths and injuries in earthquakes result from debris falling around the exteriors of commercial and public buildings. Unless you are in an especially dangerous building, such as an unreinforced brick building or a nonductile concrete-frame building, it is usually best simply to wait until the shaking has stopped before exiting the building. Then choose your exit with care, and as you leave, move calmly but quickly and watch for falling debris.

If you attempt to leave a house during the quake, be wary of collapsing chimneys or porch canopies, and watch for fallen electrical wiring. Once you are outside, or if you are outside when the quake strikes, stay well away from high buildings and masonry walls. Remain in an open area until ground motions have ceased, and do not reenter your house until you are certain that the quake is over and the walls are stable.

If you are in an automobile during the earthquake, stop in an open area away from tall buildings and overpasses, and remain in the car until the quake is over.

After an Earthquake

Afterward, the first thing to do is to check people around you for injuries. Seriously injured people should not be moved unless they are in immediate danger of further injury. First aid should be administered. Your next concern should be the danger of fire. Check the gas lines and, if there is any likelihood of leakage, turn off the main gas valve, which is usually near the meter. If gas lines outside your building may be damaged, put up warning signs, notify your neighbors and emergency authorities, and stay well clear of the fire and explosion hazard until it has been checked by a utility worker. Do not use electrical appliances if there is a possibility of gas leakage, because sparks could ignite the gas.

Check the water and electrical lines next, and turn off the main valve or breaker if lines are damaged. Also, since there is likely to be shattered glass and other debris after a quake, put on shoes before you begin to inspect the damage. There is always a danger of falling debris long after the quake, so be cautious in moving about or near any building. Particular attention should be given to the chimney.

Do not use the telephone except in a genuine emergency. Lines are always overloaded during a disaster, and your unnecessary call could block an emergency call by someone else. Use a radio to obtain information and damage reports.

Stay away from beaches and other low-lying waterfront areas where tsunamis could strike after the quake. Your radio may broadcast alerts regarding the danger of sea waves. Also stay away from steep, landslide-prone areas, since an aftershock may trigger a slide or avalanche.

Small children often suffer psychological trauma during a quake and should be comforted and attended at all times. An aftershock may panic them especially if their parents are absent, so make sure to provide them with reassurance and comfort, and watch over them until their parents arrive. Pets also suffer trauma, so attend to their needs and keep them calm.

If the water is off, emergency drinking and cooking water can be obtained from water heaters, toilet tanks, melted ice cubes, and canned vegetables. Do not flush your toilet until you are certain that you will not need the water stored in the upper tank. If the electricity is off, eat frozen, refrigerated, and fresh foods first, before they spoil. Save canned and dried food for last. Use outdoor barbecues or fireplaces for cooking, but make sure there are no nearby gas leaks before lighting any fires. In addition, before using the fireplace, check that the chimney is undamaged, especially at the roof level and in the attic.

Finally, be prepared for strong and possibly damaging aftershocks and do not remain in or near a building that has serious damage. Masonry buildings are especially susceptible to aftershock damage, and it is a good general rule to simply stay away from them until they have been declared safe by an expert.

(Figures 1 and 2) Critical infrastructure may be down following a major earthquake. This was certainly the case following the Northridge 1994 earthquake (figure 1) and the Kobe earthquake of 1995 (figure 2). Make sure you have a plan for what to do if you are separated from your family in an earthquake. Have earthquake disaster kits at home, work, and in your car.

(Figure 1)

(Figure 2)

DEVELOP A PLAN FOR YOUR FAMILY

Fewer than 10 percent of households have disaster plans. Fewer than 50 percent have disaster supply kits. What you do immediately following an earthquake may prevent injuries and significant financial loss. Preparing for an earthquake takes two forms—physical preparation (equipment and supplies) and mental preparation (knowing what to do).

State and local officials advise residents that outside help after a large, damaging quake should arrive within seventy-two hours (and we believe that the average home has, with a few simple additions, sufficient supplies and food on hand to last that long.) Immediately following an earthquake help from local fire and police departments will not be available. Be prepared to administer first aid or put out small fires. Listed below are a few suggestions.

Develop a plan for your family to use during an earthquake. Discuss it fully and conduct a drill. Draw a floor plan of your house and locate the following:

- Safest place in the house
- Most dangerous places
- Exits and alternative exits
- First-aid kit. Most homes already have the necessary items to handle routine accidents. A first-aid kit and book should be kept in a central location. Take a first-aid and CPR course from the Red Cross.
- Utility shutoff valves
- Flashlights and batteries
- Fire extinguishers. Keep one or more extinguishers on hand, and learn how to use them! Have them serviced annually to be sure they are working properly.
- Food and water supplies (72 hours). Most houses have enough simple food on hand to last several days. Eat food in your refrigerator and freezer first. If there is no power, frozen food will usually keep three days in an unopened freezer. The water heater (containing 30 to 40 gallons) for a family of four should contain enough water to last four days. The upper toilet tank can provide up to 7 gallons of water. Ice cubes in the freezer and liquid from canned food can be used, too. We recommend that you have 72 hours' worth of water available for each family member.
- Batteries and transistor radio

You should also:

- Make special provisions for elderly or disabled family members.
- Know the evacuation plans for your children's school or day care.
- Identify a person outside the immediate area who can coordinate family contact. Although local phone lines may be down, long-distance lines will function sooner.
- Gain some knowledge of first-aid procedures. Medical facilities are always overloaded after a disaster.
- Ensure that all responsible family members know what to do to avoid injury and panic. They should know the location of the main gas and water valves and electrical switch, and they should understand all safety measures that need to be taken to protect themselves, others, and the building.
- Make special provisions for pets. They will not be allowed in shelters and will need to be confined in a safe room in your house.

There is a wealth of additional guidance on how to prepare your family for an earthquake. A few suggestions are listed in Appendix D.

After an earthquake:

- Check for injuries, give first aid, and cover the seriously injured with blankets to prevent shock. Do not attempt to move seriously injured people unless they are in immediate danger.

- Turn on a battery-operated or car radio and listen for information on what to do.

- Do not attempt to drive anywhere. Roads may be damaged or blocked with debris. Freeway overpasses may be down. Road use may be restricted to emergency vehicles only.

- Plan for strong aftershocks. Stay out of already weakened and damaged homes.

- Wear shoes and protect your feet from broken glass and debris.

- Don't take unnecessary risks. If you notice any damage, turn off gas, electricity, and water. Do not turn them back on until each has been properly inspected.

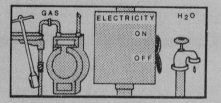

- Do not touch downed power lines or objects in contact with power lines.

- Do not expect the telephone (both landline and cellular) to work. Do not try to use the telephone unless you have an emergency.

DEVELOP A PLAN FOR YOUR BUSINESS

A comprehensive approach to disaster-mitigation, business-continuity, and crisis-management planning involves understanding your risks and their potential impacts on the business; developing risk-control, disaster-mitigation, and business-continuity and crisis-management plans; and implementing and exercising those plans.

Step 1: Risk Assessment
- Understand and define critical assets, systems, operations, people, and functions.
- Assess threats in terms of their likelihood and consequence.
- Conduct a business impact analysis, including defining time-sensitive business functions and recovery objectives.

Step 2: Disaster-Mitigation, Business-Continuity, and Crisis-Management Plan Development
- Develop prioritized risk mitigation measures, including cost-benefit and mitigation-effectiveness analyses.
- Develop business-recovery strategies for achieving recovery objectives, including the management of agreements with suppliers and distributors.
- Develop a business-continuity plan, including emergency response procedures and crisis-management and communication strategies.

Step 3: Test and Exercise the Plans
- Test and exercise your business-continuity plan. Train staff to follow such a plan.
- Coordinate with hospitals and external agencies (fire, police, and the like).
- Review the plan and make suggestions for continued improvement.
- Maintain and update plan.

Please see Appendix D for links to more information.

WHAT WILL THE GOVERNMENT DO FOR YOU?

- The primary form of disaster relief comes in the form of low-interest loans to eligible individuals, homeowners, and businesses, made available through the Small Business Administration (SBA) and intended to repair or replace damaged property and personal belongings not covered by insurance.
- The maximum SBA personal-property loan is $40,000, and the maximum SBA real-property loan for primary home repair is $200,000.
- FEMA disaster grants for emergency home repairs and temporary rental assistance are available only to households and individuals who do not qualify for loans.
- The average FEMA grant is less than $15,000. (the maximum is $26,200.) This is not enough to rebuild your home.
- The Farm Service Agency (FSA) offers loans to assist agricultural businesses.

Please see Appendix D for links to more information.

Appendix A
EARTHQUAKE HAZARD MAP OF THE UNITED STATES

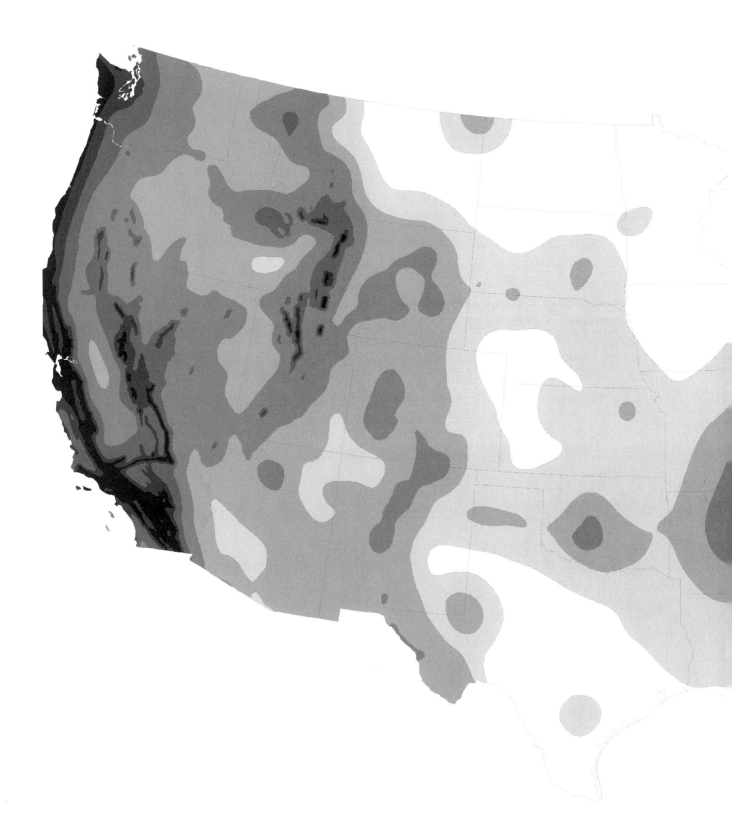

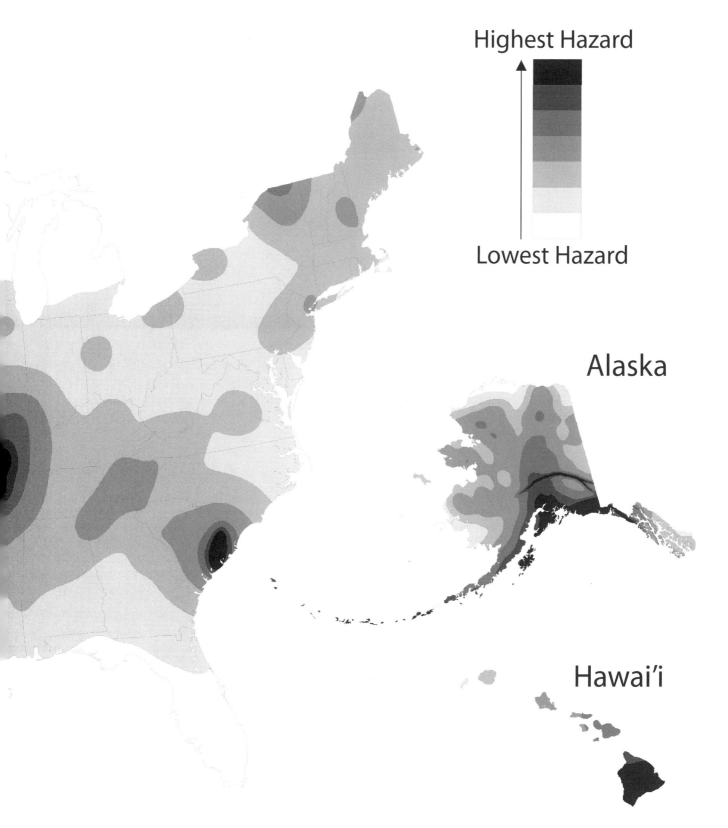

APPENDIX A

Appendix B
EARTHQUAKE HAZARD MAP OF THE WORLD

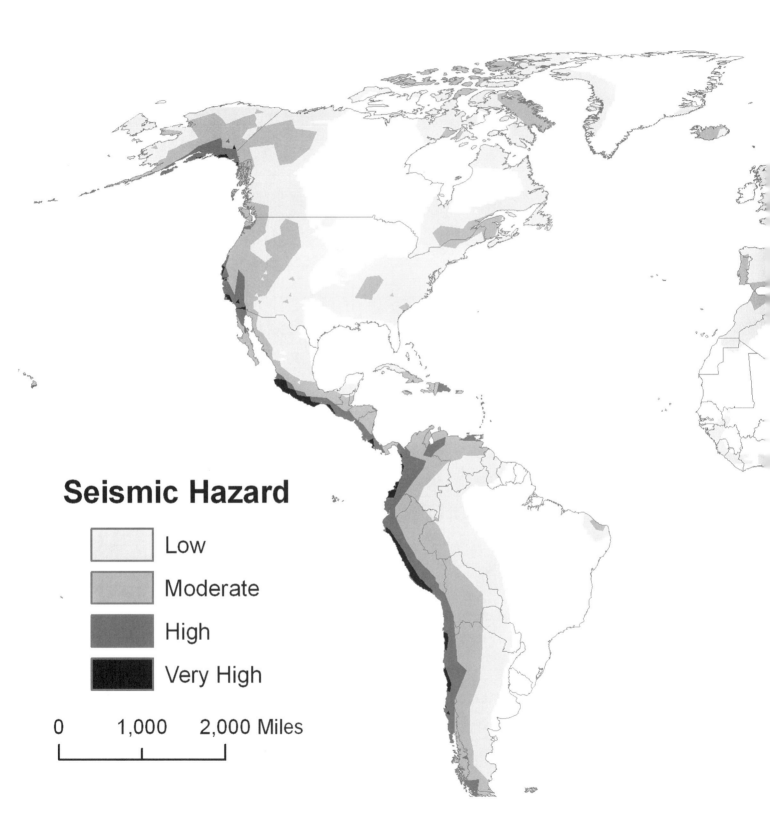

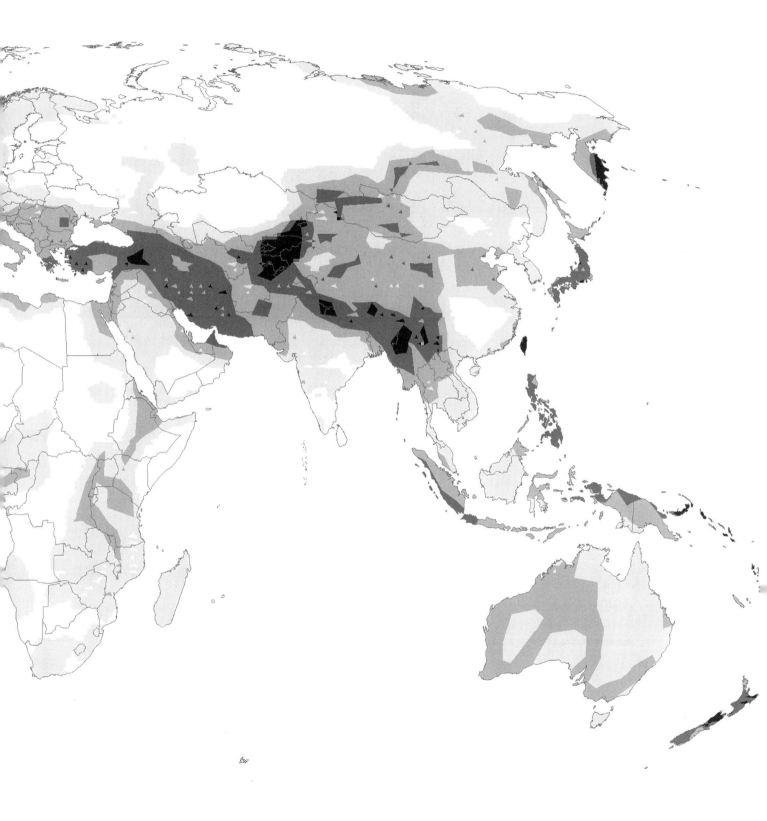

Appendix C
EARTHQUAKE INFORMATION IN THE
DIGITAL WORLD
by J. Luke Blair, United States Geological Survey

Acquiring information in the digital age is dominated by the Internet, and copious earthquake-related information and data exist within the World Wide Web. This section highlights useful sources of online information provided by the United States Geological Survey (USGS) and cooperating institutions. Because Web sites and online technologies change rapidly, search engines are the best way to track down current Web addresses. In addition to Web addresses (or URLs), keywords are provided in this section as a search alternative should a URL return no results.

Typing the keyword "earthquake" into any Web search engine will return the most popular Web sites providing earthquake information to the public. The USGS's Earthquake Hazards Program is generally acknowledged as the leading online resource for global seismic information and is the best starting point both for the novice and the experienced investigator. As part of the multiagency National Earthquake Hazards Reduction Program (NEHRP), the USGS has the lead federal responsibility to "provide notification of earthquakes in order to enhance public safety and to reduce losses through effective forecasts based on the best possible scientific information."

http://earthquake.usgs.gov/
    Keyword: "USGS Earthquake Hazards Program"

The USGS Earthquake Hazards Program Web site contains a massive collection of current and historic earthquake information, including real-time earthquake maps, educational resources for both teachers and students, scientific publications, and resources for preparedness and response. It also serves as a portal to Web sites of other earthquake research institutions, both domestic and foreign.

Global Earthquake Maps

http://earthquake.usgs.gov/eqcenter/recenteqsww/
    Keyword: "USGS earthquake maps"

Dominating the traffic to the USGS site are the real-time, global earthquake maps. Here one can examine the seismic activity around the world, with earthquake epicenters reported within minutes of the earthquake's occurrence. The clickable earthquake epicenter provides more information about that particular event including detailed location maps, magnitude, and additional geophysical data collected by the seismic instruments. USGS provides current earthquake data in multiple formats or "feeds" for importing into custom map applications, such as Google Maps, and the increasingly popular "virtual globe" programs such as Google Earth, NASA World Wind, and ESRI ArcExplorer.

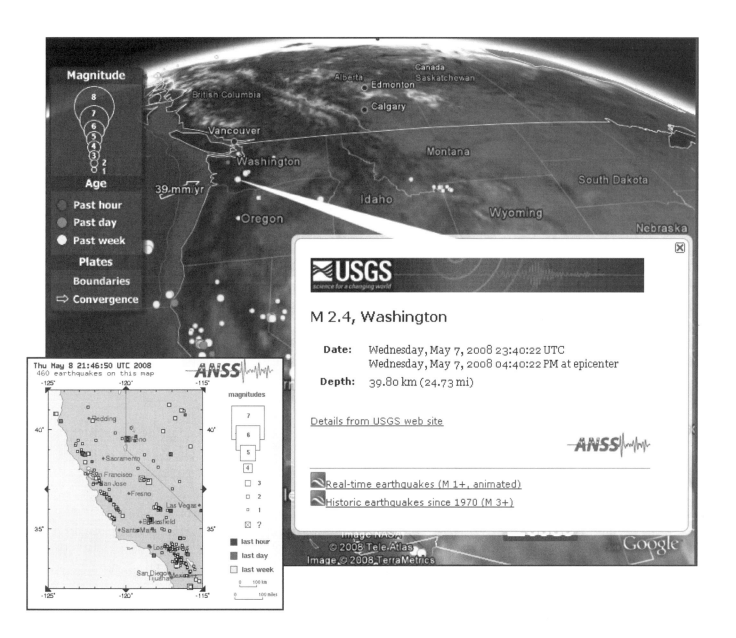

Did You Feel It?

http://earthquake.usgs.gov/dyfi/
Keyword: "USGS Did You Feel It?"

Should you experience the shaking of an earthquake, the USGS includes a "Did You Feel It?" link to a form that allows you to report the shaking and damage at your location. You can then view a map displaying accumulated data from your report and others.

Data entered through the "Did You Feel It?" form is important to earthquake scientists researching a particular event. Scientists have found that "Did You Feel It?", formally known as Community Internet Intensity (CII) reports, closely match the MMI measurements established by professional observations. In the United States, literally thousands of reports are submitted in the hours and days following an earthquake. These reports represent a dense array of human recording devices in the affected region. Shaking intensity reports often vary locally, providing scientists insight into the underlying geology as seismic waves are amplified or reduced by various rock types. The resulting maps may also be used for practical purposes, such as providing emergency response personnel with a map of those regions subjected to the strongest levels of shaking. Another benefit to collecting CII reports is to have a record of shaking in areas where seismic instruments are sparse.

ShakeMaps: USGS Real-Time Shaking Maps

http://earthquake.usgs.gov/shakemap/
Keyword: "ShakeMaps"

Following an earthquake of 3.5 or greater in California, seismic instruments recording the tremor immediately report their recorded ground-shaking data, allowing maps to be generated showing the ground motion and shaking intensity experienced in the affected area. These maps are called ShakeMaps and are released generally within 10 minutes after the earthquake event. The initial maps are typically estimates, and as seismologists examine and further process the data, increasingly accurate Shake-Maps are generated over the days and weeks following the event. ShakeMaps are used by federal, state, and local agencies for post-earthquake response and recovery, public and scientific information, and for preparedness exercises and disaster planning. Global Shake-Maps are generated for larger earthquakes (magnitude 5 or greater) but often have less detail due to scarcity of recording stations.

Virtual Globes and Applications to Earthquake Science

Moving into the third dimension, Google Earth, NASA World Wind, and ESRI ArcExplorer, are currently three popular "virtual globe" programs that promise to represent the future for examining the Earth from a home computer. In contrast to the Web site maps mentioned previously, these virtual globe programs are downloaded and installed on a home computer. Providing the same address mapping capabilities as their 2D predecessors, virtual globe programs bring in the third dimension by allowing the user to view the earth from an oblique, or "bird's-eye," perspective. High-resolution imagery draped over topography provides a realistic flying experience. The USGS is utilizing the popularity of virtual globe programs by providing compatible files (.KML files), which plot important geographical features on the globe. The ability to interact and view geologic features such as plate boundaries, earthquake faults, and their related earthquakes is very powerful for understanding just how one's residence fits within the geologic and seismic context of the globe.

Earthquake Hazards Program data in .KML format can be found here:

http://earthquake.usgs.gov/research/data/google_earth/
Keyword: "USGS Google Earth files"

U.S. Hazard Information by State

http://earthquake.usgs.gov/regional/states/
Keyword: "USGS earthquake information by state"

Earthquakes pose significant risks to a large portion of the United States. Nationwide, 75 million Americans in 39 states live in areas that may experience moderate to high shaking (see hazard map of U.S.). The USGS Earthquake Hazards Program Web site provides earthquake information for each of the 50 states and the District of Columbia. Information including historic events, those state institutions that collect or disseminate earthquake information, maps of current and historic activity, and other relevant topics.

The following section lists Web sites and products from the USGS and collaborating institutions for five high-risk regions including Alaska, the Pacific Northwest, Northern California, Southern California, and the New Madrid region of the central United States.

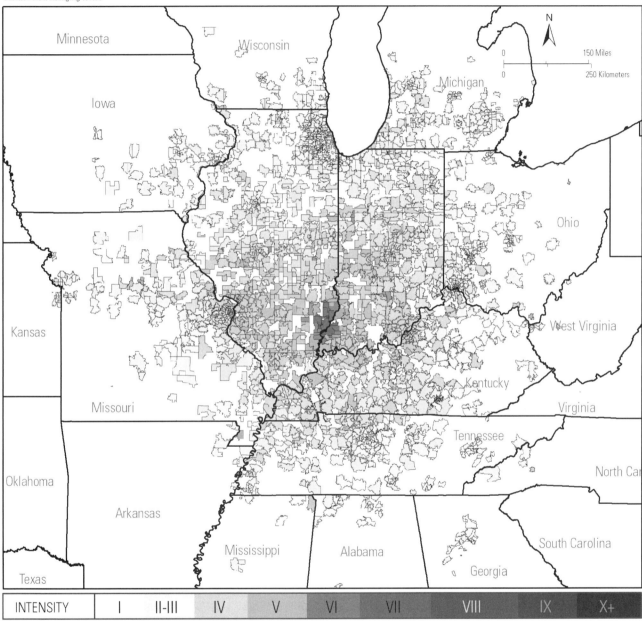

USGS "Did You Feel It?", map from the April 18, 2008, M5.2 earthquake in southern Illinois. Shaded polygons represent zip codes and average intensity reported within that area. Maximum intensity reported was VII, with reports of shaking as far away as 500 miles from the epicenter. More than 39,000 reports were logged within 24 hours of the earthquake.

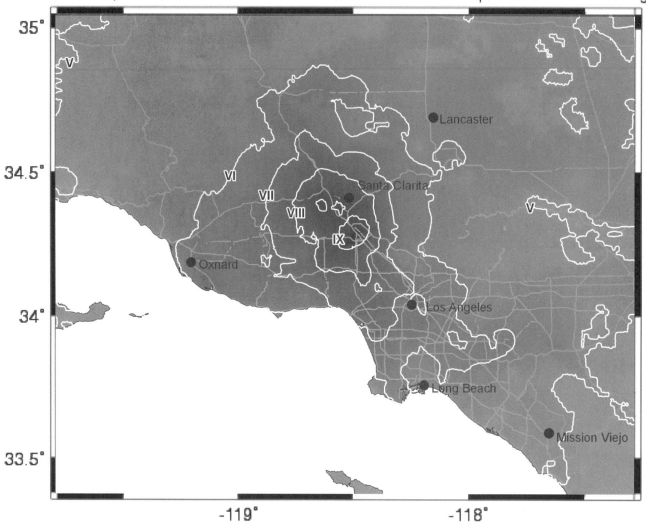

PERCEIVED SHAKING	Not felt	Weak	Light	Moderate	Strong	Very strong	Severe	Violent	Extreme
POTENTIAL DAMAGE	none	none	none	Very light	Light	Moderate	Moderate/Heavy	Heavy	Very Heavy
PEAK ACC.(%g)	<.17	.17-1.4	1.4-3.9	3.9-9.2	9.2-18	18-34	34-65	65-124	>124
PEAK VEL.(cm/s)	<0.1	0.1-1.1	1.1-3.4	3.4-8.1	8.1-16	16-31	31-60	60-116	>116
INTENSITY	I	II-III	IV	V	VI	VII	VIII	IX	X+

USGS ShakeMap for the January 17, 1994, M6.7 Northridge earthquake. Better shown in color, a first-run ShakeMap is generated with in ten minutes of an earthquake event and depicts ground-shaking intensity in the affected region.

Alaska

Alaska is the most earthquake-prone state and one of the most seismically active regions in the world, experiencing a magnitude 7 earthquake almost every year and a magnitude 8 or greater earthquake every 14 years on average. The Alaska Earthquake Information center (AEIC) is a good stop for detailed Alaska earthquake information and preparedness assistance.

http://www.aeic.alaska.edu/
Keyword: "Alaska earthquake information center"

A printable poster entitled "Earthquakes in Alaska," nicely summarizes the overall geologic setting in Alaska that produces earthquakes.

http://geopubs.wr.usgs.gov/open-file/of95-624/
Keyword: "Earthquakes in Alaska poster"

Pacific Northwest

The USGS along with state governments and universities have produced a Web site for hazards affecting the urban centers of western Washington and Oregon.

http://geomaps.wr.usgs.gov/pacnw/
Keyword: "Pacific Northwest geologic mapping and urban hazards"

The USGS has produced a new series of earthquake hazard maps for the city of Seattle. Seattle sits atop a sedimentary basin that strongly affects the patterns of earthquake ground shaking and, therefore, of potential damage. Detailed urban seismic hazard maps provide a high-resolution view of the potential for strong earthquake shaking in Seattle.

http://earthquake.usgs.gov/regional/pacnw/hazmap/seattle/
Keyword: "Seattle seismic hazard maps"

Northern California

Developed by the USGS and a large group of organizations, "Putting Down Roots in Earthquake Country" is a handbook providing information about the threat posed by earthquakes in the San Francisco Bay region and explains how you can prepare for, survive, and recover from inevitable earthquake events.

http://pubs.usgs.gov/gip/2005/15/
Keyword: "Northern California Putting Down Roots"

The Association of Bay Area Governments (ABAG), the nine-county regional planning and services agency for the Bay Area, has a comprehensive website with everything from disaster scenario models, liquefaction susceptibility maps, and tips for making your home safer.

http://quake.abag.ca.gov/
Keyword: "ABAG Earthquake Maps and information"

Southern California

The Southern California Earthquake Center (SCEC) is a community of over six hundred scientists, students, and others at over sixty institutions worldwide, headquartered at the University of Southern California. SCEC is funded by the National Science Foundation and the U.S. Geological Survey to develop a comprehensive understanding of earthquakes in Southern California and elsewhere and to communicate useful knowledge for reducing earthquake risk.

http://www.scec.org/
Keyword: "Southern California Earthquake Center"

Earthquake professionals, business and community leaders, emergency managers, and others have joined together to organize Dare to Prepare, an earthquake readiness campaign to raise earthquake awareness and encourage earthquake readiness in Southern California. The campaign is supported with funding from federal, state, and local partners.

http://www.daretoprepare.org/
Keyword: "Southern California dare to prepare"

New Madrid Region, Central United States

The Center for Earthquake Research and Information (CERI), is a highly regarded institute run out of the University of Memphis, and a source of earthquake data, information, and preparedness resources.

Index

A

Adobe buildings, 109
Aftershocks, 31, 167
Air conditioners, 150
Alaska
    1964 earthquake, 33, 63, 64–65, 89
    earthquake frequency in, 183
    earthquake hazard map of, 175
    earthquake information for, 183
    tsunamis and, 61, 63
Alluvial soils, 58
Alpide belt, 22
Alquist-Priolo (AP) zones, 39, 43
Anchorage, Alaska, 64–65
Anheuser-Busch, 18–19
Architectural ornaments, 133
Association of Bay Area Governments (ABAG), 183

B

Bakersfield earthquake (1952), 104
Balconies, 133
Baldwin Hills Reservoir, 69, 71
Base isolation, 142–43
Bay of Plenty, New Zealand, earthquake (1987), 28, 36, 47
Beds, location of, 151
Berkeley, California, 28, 79, 108, 140, 143
Bhuj, India, 23, 108
Big Bear earthquake (1992), 135
Bookcases, 151, 153
Bracing, 78–79, 83–85
Brick veneer, 87–88
Building codes, 39, 75, 100, 102, 138–39
Buildings
    adjacent, 72–73
    adobe, 109
    basic structural components of, 77
    cavity-wall brick, 109
    clay-tile, 90
    concrete-block, 89
    concrete-frame, 96–101
    concrete shear-wall, 94–95
    concrete tilt-up, 101–3
    damage to, from faulting, 40–41
    earthquake resistance in, 75–79, 111, 142
    effects of earthquake forces on, 76–77
    foundations for, 112–22
    with heavy roofs, 128–29
    irregularly shaped, 130–31
    old, 100, 117, 122, 136–37
    reinforced masonry, 90
    resonance and, 54
    retrofitting, 140
    "rocking," 141
    rubble, 109
    safety rankings of, 109
    soft-story, 100, 123–25, 126, 128
    split-level, 123, 124
    steel-framed, 78, 91–93
    on stilts or pilings, 122, 123
    stone, 109
    tall concrete, 75, 95
    unreinforced masonry (URM), 104–8
    wood columns in, 123
    wood-frame, 83–85, 114–15, 117, 118–21
    wood-frame, with masonry veneer, 87–88
    wood-frame, with stuccoed walls, 86
Businesses
    avoiding interruption of, 154–57
    disaster plans for, 172
    earthquake insurance for, 161
    earthquake risk management for, 17–19

C

Cabinets, 151, 153
Calaveras fault, 25, 27, 42, 88
California. See also individual cities and earthquakes
    dams in, 69, 70
    earthquake information for, 183
    earthquake insurance in, 161, 162–63, 164–65
    faults in, 30, 42
    major earthquakes in, 30
    probability of major earthquake for, 45
    unreinforced masonry structures in, 106, 108
California Earthquake Authority (CEA), 163, 164–65
Cape Ann earthquake (1755), 21
Cascadia subduction zone, 25, 63
Cavity-wall brick buildings, 109
Central Luzon, Philippines, earthquake (1990), 54, 96
Chile earthquake (1960), 24
Chimneys, 134–36

Circum-Pacific seismic belt, 22
Clay-tile buildings, 90
Cliffs, 59, 60
Coalinga earthquake (1983), 26, 30, 42, 88, 104, 137, 163
Columns, wood, 123
Community Internet Intensity (CII) reports, 180
Compton, California, 104
Concord fault, 88
Concrete-block buildings, 89
Concrete-frame buildings, 96–101
Concrete shear-wall buildings, 94–95
Concrete tilt-up buildings, 101–3
Continental drift, 22–24
Crawl-space walls, sheathing, 115, 120–21
Crescent City, California, 63

D

Daly City, California, 42, 57, 59, 60
Dams, 69–70
Diagonal bracing, 78, 79, 83–84, 85
Diaphragm connections, 128–29
Dikes, 71
Disaster plans, 170–72
Disaster relief, 173
Doors, large, 132
Dual System, 91

E

Earthquake hazard maps
    of the United States, 174–75
    of the world, 176–77
Earthquake insurance
    availability of, 161
    in California, 161, 162–63, 164–65
    commercial, 161
    cost of, 161, 162
    coverage provided by, 162
    deductibles for, 161–62
    overview of, 161–62
    purchase decision for, 15–16, 163
Earthquake Risk Management program
    for businesses, 17–19
    for homes, 13–16
Earthquakes (general)
    actions after, 167–68, 171
    aftershocks following, 31, 168
    causes of, 21–25, 26–27
    deaths from, 26, 72, 104, 133
    displacement of ground by, 27
    effects of, on buildings, 76–77
    epicenters of, 32
    experiencing strong, 37
    fault creep and, 27
    fires following, 147
    foreshocks preceding, 31
    hypocenters of, 32
    information on, by state, 180, 183
    injuries during, 167
    measuring, 32–35
    preparing for, 167, 170
    safety during, 167, 170
    shallow- vs. deep-focus, 24
    websites about, 178
    worldwide relative density of, 22, 23
Earthquakes (specific). See also Kobe, Japan, earthquake (1995);
    Loma Prieta earthquake (1989); Northridge earthquake (1994);
    San Fernando Valley earthquake (1971);
    San Francisco earthquake (1906)
    1755 Cape Ann, 21
    1755 Lisbon, Portugal, 21
    1812 New Madrid, 35
    1836 Hayward, 46
    1868 Hayward, 46
    1872 Owens Valley, 27
    1925 Santa Barbara, 69
    1933 Long Beach, 58, 69, 100, 104, 106, 136, 137
    1952 Bakersfield, 104
    1952 Kern County, 31
    1957 San Francisco, 59, 60
    1959 Hebgen Lake, Montana, 27
    1960 Chile, 24
    1964 Alaska, 33, 63, 64–65, 89
    1964 Niigata, Japan, 54, 55
    1969 Santa Rosa, 136
    1972 Managua, Nicaragua, 88, 128
    1976 Friuli, Italy, 107
    1978 Sendai, Japan, 93
    1983 Coalinga, 26, 30, 42, 88, 104, 137, 163
    1984 Morgan Hill, 117, 132, 152
    1985 Mexico City, 54, 98
    1985 Santiago, Chile, 124
    1987 Bay of Plenty, New Zealand, 28, 36, 47
    1987 Whittier Narrows, 26, 42, 104, 132, 150, 163

1990 Central Luzon, Philippines, 54, 96

1992 Big Bear, 135

1993 Hokkaido, Japan, 50, 62

1999 Taiwan, 71, 94, 99

1999 Turkey, 28

2001 Gujarat, India, 23, 108, 109

2001 Nisqually, Washington, 24, 25, 58, 106

2004 Indonesia, 24, 33, 60, 61, 63

2004 Niigata, Japan, 36

2004 San Simeon, 133

2005 Pakistan, 23

2007 Niigata, Japan, 54, 55, 98, 129, 156, 168

2008 Wells, Nevada, 104

2008 Wenchuan, China, 23, 74, 81, 109

East Bench fault zone, 42

Elastic rebound, 24

Electrical shutoff, 148

Engineers, licensed, 113

Epicenters, 32

Equipment, minimizing damage to, 156

F

Fault creep, 25, 27

Faulting
    building damage caused by, 40–41
    definition of, 27
    direction of, 24
    lateral, 41
    vertical, 40

Faults
    definition of, 26
    location of, 42–43
    movement of, 24–25, 26–27, 28
    plate tectonics and, 23
    sag ponds along, 71
    types of, 24–25
    underrated, 48–49
    unmapped, 42–43

Fault systems, 26

Fault zones
    definition of, 26
    distance from, 47
    hazards of, 39
    property in, 39–40, 47

FEMA disaster grants, 173

Fences, 152

Fire, 147

Fireplaces, 134–36

First-aid kits, 170

Food and water supplies, 170

Foreshocks, 31

Foundations
    connections to, for wood-frame buildings, 114, 115, 117, 118–19
    cracks in, 122
    existing damage to, 122
    importance of, 112
    on rock, 53
    types of, 112

Frame-action bracing, 79

Friuli, Italy, earthquake (1976), 107

Furnaces, 150

Furnishings, 151, 153

G

Garden walls, 152

Gas
    leaking, 145
    shutoff, 148

Geologists, 113

Glassware, 151

Government assistance, 173

Gujarat, India, earthquake (2001), 109

Gutenberg, Beno Francis, 33

Gutenberg-Richter scale, 33

Gypsum board, 150

H

Hanshin Expressway, 97

Hawaii
    earthquake hazard map of, 175
    tsunamis and, 61

Hayward fault, 27, 40, 42, 46, 71

Healdsburg fault, 46

Hebgen Dam, 70

Hebgen Lake, Montana, earthquake (1959), 27

Help, seeking, 113

Highway construction, 96, 97

Hillsides, 59, 64–65, 66

Himalayas, 23

Hokkaido, Japan, earthquake (1993), 50, 62

"Hold-down" connectors, 103

Hollister, California, 25, 27, 32, 103

Houses. See Buildings

Hurricane Katrina, 71
Hypocenters, 32

I

Imperial fault, 42
India
    2001 earthquake, 23, 108, 109
    continental drift and, 23
Indonesia earthquake (2004), 24, 33, 60, 61, 63
Insurance. See Earthquake insurance
Intensity maps, 34-35
Interior damage, minimizing, 145, 150-51, 152, 153
International Building Code (IBC), 100

K

Kern County earthquake (1952), 31
Kern River fault, 25
Kobe, Japan, earthquake (1995)
    collapsed and damaged buildings from, 72, 80, 91, 94, 128, 138
    freeway collapse from, 97, 168
    losses from, 44, 163
    port damaged by, 55, 57

L

Landers fault, 28
Landfills, 51, 55, 57
Landslides, 64-67
Lateral bracing, 78
Levees, 71
Liquefaction, 54, 55, 57
Lisbon, Portugal, earthquake (1755), 21
Local magnitude scale, 33
Loma Prieta earthquake (1989)
    aftershocks following, 31
    collapsed and damaged buildings from, 51, 53, 55, 59, 60, 72, 73, 83, 87, 90, 92, 96, 103, 104, 107, 114, 115, 117, 124, 125, 132, 160
    deaths from, 72, 104
    focus of, 24
    intensity map for, 34, 35
    landslides and, 64
    losses from, 163
Long Beach earthquake (1933), 58, 69, 100, 104, 106, 136, 137
Los Angeles, 42, 45, 58, 64. See also individual quakes

M

Managua, Nicaragua, earthquake (1972), 89, 128
Masonry veneer, 87-88
Memphis, Tennessee, 42, 48

Mexico City earthquake (1985), 54, 98
Mid-Atlantic Ridge, 22, 24
Modified Mercalli Intensity (MMI) scale, 33-35
Moment Magnitude scale, 33, 34
Morgan Hill earthquake (1984), 117, 132, 152

N

National Earthquake Hazards Reduction Program (NEHRP), 178
New Madrid, Missouri
    1812 earthquake, 35, 139
    buildings in, 139
    earthquake information for, 183
New Orleans, 71
Newport-Inglewood fault, 42, 71
Niigata, Japan
    1964 earthquake, 54, 55
    2004 earthquake, 36
    2007 earthquake, 54, 55, 98, 129, 156, 168
Nisqually, Washington, earthquake (2001), 24, 25, 58, 106
North Anatolia fault, 28
Northridge earthquake (1994)
    aftershocks following, 31
    cause of, 22
    collapsed and damaged buildings from, 18-19, 53, 83, 91, 92, 95, 96, 97, 99, 100, 101, 102, 106, 114, 115, 126, 160
    deaths from, 26
    fire following, 147
    focus of, 24
    free collapse from, 168
    freeways and, 97
    landslides and, 64, 65
    losses from, 161, 163
    MMI map for, 35
    ShakeMap for, 182
    on unmapped fault, 42

O

Oakland, California, 42, 53, 58, 90, 92, 96, 108
Olive View Hospital, 94-95
    1000 Island Lake, 22
Orange County, California, 101
Oriented Strand Board (OSB), 83
Owens Valley earthquake (1872), 27

P

Pacific Northwest. See also individual cities
    earthquake information for, 183
    future earthquakes in, 25

Pakistan earthquake (2005), 23
Palmdale, Lake, 71
Parapets, 132, 133
Pictures, hanging, 153
Pilings, buildings on, 122, 123
Plaster, 150
Plate tectonics, 22-24
Pleito fault, 25
Plywood shear-wall bracing, 78, 79, 83-84, 85, 128, 132
Port facilities, 58
Portland, Oregon, 48, 49
Puerto Rico, 61
Puget Sound area, 24, 25, 58, 101

R
Reinforced masonry buildings, 90
Reservoirs, 69-70
Resonance, 54
Retaining walls, 152
Retrofitting, 140
Richter, Charles E., 33
Richter scale, 33
Ridges, 59
Rock foundations, 53
Rodgers Creek fault, 46
Roofs
    heavy, 128-29
    strengthening, 129
Rose Canyon fault, 42
Rubble construction, 108, 109

S
Sacramento, California, 140
Sacramento–San Joaquin Delta, 71
Sag ponds, 71
St. Louis, Missouri, 48, 109, 139
Salt Lake City, Utah, 39, 42, 48, 49, 57, 101, 109, 143
San Andreas fault
    appearance of, 26
    area covered by, 26
    development along, 42
    movement of, 25, 26, 27, 41
    plate tectonics and, 22, 23, 24
    reservoirs along, 71
San Diego, California, 42, 48, 58
San Fernando fault, 25

San Fernando Valley earthquake (1971)
    aftershocks following, 31
    collapsed and damaged buildings from, 26, 39, 40, 88, 89, 90, 94, 101, 102, 117, 124, 133, 134, 135, 152
    dam failure from, 69
    deaths from, 26
    landslides and, 64
    losses from, 163
    unexpectedness of, 26, 42
San Francisco earthquake (1906)
    1957 earthquake, 59, 60
    accounts of, 37, 51
    building codes in, 136
    chimneys destroyed in, 134
    collapsed and damaged buildings from, 53, 58, 80, 114, 117
    epicenter of, 32
    liquefaction and, 55, 57
    fire following, 136, 147
    intensity of, 33, 34, 35, 36, 53
    rebuilding after, 136
    retrofitting in, 140
San Andreas fault and, 42
    unreinforced masonry structures in, 106, 108
San Francisco Bay Area
    earthquake information for, 183
    estimated consequences of major earthquake in, 44
    faults in, 45, 46
    probability of major earthquake for, 44
San Francisco–Oakland Bay Bridge, 93
San Gabriel Mountains, 26
San Jose, California, 42, 53, 108
San Simeon earthquake (2004), 133
Santa Barbara, California
    1925 earthquake, 69
Santa Ynez fault zone and, 42
Santa Cruz, California, 64, 72, 104, 117
Santa Rosa, California, 53, 136, 137, 146
Santa Ynez fault zone, 42
Santiago, Chile, earthquake (1985), 124
Seattle, Washington, 25, 39, 42, 48, 49, 57, 58, 75, 139, 183
Seiches, 70
Sendai, Japan, earthquake (1978), 93
Shear-wall bracing, 78, 79, 83-84, 85, 128, 132
Sheetrock, 150
Sheffield Dam, 69
Shelves, 153

Shihkang Dam, 71
Sierra Nevada, 22, 26
Silicon Valley, California, 46, 101
Small Business Administration (SBA), 173
Soft-story designs, 100, 123–25, 126, 128
Soil
    liquefaction, 54, 55, 57
    risk and, 51, 58
Southern California Earthquake Center (SCEC), 183
Spitak, Soviet Armenia, 93
Split-level houses, 123, 124
Steel-framed buildings, 78, 91–93
Stilts, buildings on, 122, 123
Stone buildings, 109
Stone veneer, 87–88
Structural Engineers Association (SEA), 113
Stucco buildings, 86
Swimming pools, 152

T
Taiwan earthquake (1999), 71, 94, 99
Temescal, Lake, 71
Tension cracks, 41
Tilt-up buildings, 101–3
Tokyo, Japan, 141
Tsunamis, 60, 61–63
Turkey earthquake (1999), 28
Turnagain Heights slide, 64–65

U
Una Lake, 71
Uniform Building Code (UBC), 100, 102
United States, earthquake hazard map of, 174–75
United States Geological Survey (USGS)
    "Did You Feel It?", 180, 181
    Earthquake Hazards Program, 178, 180
    global earthquake maps, 178
    intensity maps, 34
    ShakeMaps, 180, 182
Unreinforced masonry (URM) buildings, 104–8
Utilities
    minimizing damage to, 145–46
    shutting off, 148

V
Vajont reservoir, 70
Vancouver, British Columbia, Canada, 58
Van Norman Dam, 68, 69

"Virtual globe" programs, 178, 180

W
Wallboard, 150
Wasatch fault, 25, 42
Water
    damage, 145
    heaters, 149, 150
    shutoff, 148
    sites near, 58–59
    supply of, 170
Watsonville, California, 117, 146
Wells, Nevada, earthquake (2008), 104
Wenchuan, China, earthquake (2008), 23, 74, 81, 109
White Wolf fault, 25, 42
Whittier Narrows earthquake (1987), 26, 42, 104, 132, 150, 163
Windows, large, 132
Wine, protecting, 158
Wood diagonal bracing, 83–84, 85
Wood-frame buildings, 83–85
    foundation connections for, 114, 115, 117, 118–19
    with masonry veneer, 87–88
    sheathing crawl-space walls for, 115, 120–21
    with stuccoed walls, 86

X
X-bracing, 78, 79

PHOTOGRAPHIC, MAP, AND ILLUSTRATION CREDITS

Introduction
Figure 3: Courtesy David Strykowski

Chapter 1
Figure 1: After D. L. Anderson
Figure 3: Created by J. Luke Blair, US Geological Survey, USGS information
Figure 8: After D. L. Anderson
Figure 9: U.S. Geological Survey
Figure 12: Robert L Murray
Figure 20: Created by J. Luke Blair, US Geological Survey, USGS information
Figure 26: Created by J. Luke Blair, US Geological Survey, USGS information
Figure 27: Created by J. Luke Blair, US Geological Survey, USGS information
Figure 28: Created by J. Luke Blair, US Geological Survey, USGS information
Figure 29: Created by J. Luke Blair, US Geological Survey, USGS information
Figure 30: US Geological Survey
Figure 31: Earthquake Engineering Research Institute
Page 20: Bear Photo Collection (Courtesy K.V. Steinbrugge)
Page 25, Sidebar: Courtesy Mark Pierepiekarz

Chapter 2
Figure 1: Created by J. Luke Blair, US Geological Survey, USGS information
Figure 2: Created by J. Luke Blair, US Geological Survey, USGS information
Figure 3: U.S. Geological Survey
Figure 4: U.S. Geological Survey
Figure 5: Created by J. Luke Blair, US Geological Survey, USGS information, ESRI base map data
Page 40: U.S. Geological Survey
Page 41, upper left: Carnegie Institution
Page 41, upper: U.S. Geological Survey
Page 41, lower: U.S. Geological Survey
Page 44, Sidebar: Created by J. Luke Blair, US Geological Survey, USGS information
Page 45, (bottom image): Produced by J. Luke Blair, US Geological Survey, USGS information; Top charts: Field, Edward H., Milner, Kevin R., and the 2007 Working Group on California Earthquake Probabilities, U.S. Geological Survey, Fact Sheet 2008-3027
Page 46, Sidebar: Created by J. Luke Blair, US Geological Survey, USGS information
Pages 48-49, Sidebar: Created by J. Luke Blair, US Geological Survey, USGS information, ESRI base map data

Chapter 3
Figure 2: Bancroft Library, University of California, Berkeley
Figure 5: Reproduced from Tobriner, Stephen: Earthquake Spectra April 2006. Original by Harry O. Wood and redrawn in black & white by John R. Freeman
Figure 6: J. Penzien (courtesy H. B. Seed)
Figure 9: Utah Geological Society
Figure 10: Created by J. Luke Blair, US Geological Survey, USGS information, ESRI base map data
Figure 17: Created by J. Luke Blair, US Geological Survey, USGS information, ESRI base map data
Figure 18: G. Griggs
Figure 19: Earthquake Engineering Research Institute
Figure 20: Earthquake Engineering Research Institute
Figure 24: U.S. Army
Figure 25: NOAA

Chapter 4:
Figure 1: Created by J. Luke Blair, US Geological Survey, USGS information, ESRI base map data
Figure 2: California Department of Water Resources
Figure 4: U.S. Army
Figure 10: *San Francisco Examiner*, Courtesy K. V. Steinbrugge
Page 68: Los Angeles Department of Water and Power

Chapter 5
Figure 1: Created by Dan Clark, ReadyDemo
Page 74: Courtesy Kit Miyamotyo
Page 77, Sidebar: Created by Dan Clark, ReadyDemo
Page 78, Sidebar (right 3 images): Created by Dan Clark, ReadyDemo

Chapter 6
Figure 1: Courtesy David Strykowski
Figure 2: Courtesy R.S. Honeyman, Jr
Figure 4: Created by Dan Clark, ReadyDemo
Figure 6: Courtesy G. & E. Y. Jelensky

Figure 7: Courtesy G. & E. Y. Jelensky

Figure 8: Courtesy R. Kiprov

Figure 10: American Plywood Association

Figure 11: Created by Dan Clark, ReadyDemo

Figure 16: California Division of Highways

Figure 17: NOAA

Figure 18: F.E. McClure

Figure 19: Pacific Gas and Electric Company

Figure 21: National Bureau of Standards

Figure 22: California Institute of Technology, Earthquake Engineering Research Laboratory

Figure 24: Courtesy David Strykowski

Figure 25: Courtesy David Strykowski

Figure 30: San Francisco Examiner, Copyright (Paul Chinn)

Figure 34: R. Augenstein

Figure 46: Renee G. Lee, Arup

Figure 60: Created by Dan Clark, ReadyDemo

Figure 62: Courtesy R.S. Honeyman, Jr.

Figure 63: Courtesy R.S. Honeyman, Jr.

Figure 68: Robert Philbrick

Page 81: Courtesy Kit Miyamoto

Chapter 7

Figure 2: Created by Dan Clark, ReadyDemo

Figure 3: Created by Dan Clark, ReadyDemo

Figure 4: Created by Dan Clark, ReadyDemo

Figure 5: Created by Dan Clark, ReadyDemo

Figure 6: N. Cery

Figure 7: Courtesy H.J. Degenkolb

Figure 19: American Plywood Association

Figure 20: Los Angeles City Department of Building and Safety

Figure 23: California Institute of Technology, Earthquake Engineering Research Laboratory

Figure 24: California Institute of Technology, Earthquake Engineering Research Laboratory

Figure 32: U.S. Army

Figure 35: Renee G. Lee, Arup

Figure 38: Courtesy R. Kiprov

Figure 41: National Bureau of Standards

Figure 46: Los Angeles City Department of Building and Safety

Figure 48: Courtesy R.S. Honeyman, Jr.

Figure 50: H.M. Engle

Figure 51: F.E. McClure

Page 110: Courtesy David Strykowski

Pages 118-119 Sidebar (all pictures): Created by Dan Clark, ReadyDemo

Pages 120-121 Sidebar (all pictures): Created by Dan Clark, ReadyDemo

Page 126, Sidebar (schematic): Created by Dan Clark, ReadyDemo

Pages 130–131, Sidebar: Created by Dan Clark, ReadyDemo

Page 139, Sidebar (top): Courtesy Mark Pierepiekarz

Page 140, Sidebar (top): Courtesy Kit Miyamoto, Miyamoto International

Page 141, Sidebar (left): Courtesy and copyright Arup, photo by Michel Denance

Page 141, Sidebar (right): Courtesy and copyright Arup, photo by Frank la Riviere

Chapter 8

Figure 2: F.E. McClure

Figure 9: P.E. Estes

Figure 10: P.E. Estes

Figure 12: Los Angeles City Department of Building and Safety

Page 144: Los Angeles Department of Water and Power

Page 144 (bottom): Renee G. Lee, Arup

Page 158, Sidebar (top left 3 images): Tom Chan

Page 159, Sidebar (bottom): Tom Chan

Page 160, top: Nick Miloslavich

Page 166: Renee G. Lee, Arup

Appendix A: Data from USGS

Appendix B: Created by J. Luke Blair, US Geological Survey, based on Global Seismic Hazard Assessment Program (GSHAP) data

All other photographs are from the authors' collections or are in the public domain.

REFERENCES AND ADDITIONAL INFORMATION

Please visit: www.peaceofmindineqcountry.com for the references used in the development of this book, as well as a variety of helpful links and resources.

Peace of mind in earthquake country : how to save your home, business, and life / Peter I. Yanev and Andrew C.T. Thompson

TH1095 .Y36 2008
Gen Stx